Ecological Studies, Vol. 129

Analysis and Synthesis

Edited by

M.M. Caldwell, Logan, USA
G. Heldmaier, Marburg, Germany
O.L. Lange, Würzburg, Germany
H.A. Mooney, Stanford, USA
E.-D. Schulze, Bayreuth, Germany
U. Sommer, Kiel, Germany

Ecological Studies

Volumes published since 1992 are listed at the end of this book.

Springer

Berlin
Heidelberg
New York
Barcelona
Budapest
Hong Kong
London
Milan
Paris
Santa Clara
Singapore
Tokyo

Tom Andersen

Pelagic Nutrient Cycles

Herbivores as Sources and Sinks

With 90 Figures and 16 Tables

 Springer

Dr. Tom Andersen
Department of Biology
University of Oslo
P.O. Box 1069
Blindern
0316 Oslo, Norway

ISSN 0070-8356
ISBN 3-540-61881-3 Springer-Verlag Berlin Heidelberg New York

Library of Congress Cataloging-in-Publication Data
Andersen, Tom, 1955-
Pelagic nutrient cycles : herbivores as sources and sinks / Tom
Andersen.
p. cm. – (Ecological studies, ISSN 070-8356 ; v. 129)
Includes bibliograhical references (p.   ) and index.
ISBN 3-540-61881-3 (hc)
1. Nutrient cycles. 2. Food chains (Ecology) 3. Herbivores–
Food. I. Title. II. Series.
QH344.A53 1997
577.7 ´16–dc21   96-49478

Cover Design: D & P Heidelberg
Typesetting: Camera ready by Dr. Kurt Darms, Bevern
SPIN 10533576   31/3137 5 4 3 2 1 0 - Printed on acid free paper -

Foreword

While ecology is one of the scientific disiplines that most clearly belongs to "basic research", it also strives to serve as a predictive tool for management. Outstanding examples of predictive ecology are Vollenweider's models on the relationship between phosphorus load and water renewal time of lakes, and the resulting algal biomass. The needed few and easily accessible input parameters to very simple models provided a direct link from basic ecology to management, and today these models are key tools for managers worldwide to control lake eutrophication and algal blooms.

The baseline of this success is the general relation between phosphorus concentration and phytoplankton biomass that is observed for most lakes. While these relationships are most frequently presented in log-log diagrams, the aquatic ecologist who replots these a linear scale may ask himself why, in spite of the overall correlation, there is still such a variability. It is possible to predict levels of algal biomass that may be synthesized at a given phosphorus load. Some lakes apparently offer optimal conditions for their phytoplankton communities, while others may support less than half the biomass at the same phosphorus load. There are limits to growth, however, and the predictive outcome of phosphorus load decreases as the load increases. This holds in particular for the higher trophic levels like zooplankton and fish. With increased enrichment there are signs of decreased efficiency of energy transfer between foodweb compartments. Moreover, system stability tends to decrease with increased load, and beyond some undefined, probably highly site-specific limits, further enrichment may cause decreased productivity at higher trophic levels. This "paradox of enrichment", as Rosenzweig (1971) labelled it, is a central theme of this book. To resolve these matters, we need to go beyond large-scale predictive models, and examine in more detail foodweb dynamics and its relationship to nutrient cycling and ecosystem stability. First of all, we need to know some of the basic, ecological features of the organisms in question. Predictive modelling without reliable inputs would be like constructing a city plan without knowing the basic demands of the people.

Daphnia is the highly appreciated key component in freshwater ecosystems. This tiny crustacean water flea is instrumental for grazing, nutrient recycling and transfer of energy to higher trophic levels. It is the obvious

choice for modelling purposes, not only because it is a good mediator of high
trophic efficiency, but also because there is a long tradition of ecophysiologi-
cal studies on these organisms that provide reliable input parameters. In
logical model development, the foundation in terms of the basic lifeand
death parameters for the central actors, *Daphnia*, algae, and bacteria, is set
before the first act is played. The main theme is then elaborated further
through increasing model complexity in the subsequent acts.

The play metaphor is in the spirit of A. J. Lotka, who in his pioneering
work from 1925, first attempted to merge nutrient cycling and stoichiometric
arguments into basic ecology. Actors and the scene as well are made of the
same stuff that recycle over and over again in the great theatre of life. This
early attempt was largely neglected, due partly to purely practical reasons. It
this book, however, the author pays an enlightening revisit to Lotka's ideas by
incorporating stoichiometric concepts in his models, clearly demonstrating
how vital they are to productivity and foodweb structure. Recent models
incorporating stoichiometric considerations in food-web studies show that
the transfer of energy in terms of carbon from autotrophs to higher trophic
levels may be governed by food quality in terms of element ratios. When food
is high in C relative to other essential elements like nitrogen and phosphorus,
excess C simply "goes to waste". The application of the Liebig minimum principle, also for the heterotrophs, is a new way of predicting "trophic efficiency"
in food webs. In a prey-predator scheme, these stoichiometric applications,
when isocline analysis is carried out, show a new, paradoxically stable state
with zero predator (grazer) biomass at high algal biomass, when the algae are
too P-deprived to support zooplankton growth. Such qualitative predator
defences should attract attention from all those working with prey-predator
systems.

However, intriguing they may be to a theoretical ecologist, we should, as
always, ask whether these models have any bearing on predictions at the
ecosystem level. They have indeed. In this book, these aspects are explored,
not merely as a performance of single actors, but as an examination of the
play in a more complete context. With a stepwise logical argumentation for
parameter inputs, the author creates a toolbox of submodels, allowing him to
judge Rosenzweig's axiom on less subjective grounds than most other people.
The data do indeed provide evidence for reduced grazer (*Daphnia*) success,
instability and chaotic ecosystem behaviour beyond a certain enrichment
threshold; enrichment thus decreases predictability. There is a lesson to be
learned from these exercises, not only for aquatic ecologists but for all those
dealing with herbivore impact on ecosystems, and the impact of enrichment.
Through a clear line of reasoning, we are guided on a very visible path all the
way from large-scale processes such as phosphorus loading, lake hydrology
and dilution rate, via species' physiology, to cellular processes and back to a
better understanding of other large-scale processes like foodweb "efficiency"
and ecosystem stability.

Oslo, January 1997   Dag O. Hessen

Preface

> *How can the tangled bank, in all its glorious, muddy complexity, be reduced to a set of antiseptic differential equations? To be fair, it isn't certain that it can; but to presume that we cannot hope to find general rules for the way nature works is to reduce every ecological interaction to a special case, to be marvelled at and then added to the vast rummage box of wonderful anecdotes and special cases.*
>
> John Lawton

Public concern about problems related to the eutrophication of inland waters reached a peak level in Norway in 1976, following the nuisance bloom of *Oscillatoria* in lake Mjøsa. Following the massive effort toward nutrient load abatement in this lake, an increasing need was felt for having better tools to determine what levels of nutrient input a given lake could tolerate without showing negative effects of eutrophication. Recent research at that time had indicated that much of the variability between lakes receiving similar nutrient loadings could be attributed to properties of pelagic food webs (Shapiro et al. 1975). Consequently, in order to increase the understanding of biological mechanisms involved in the eutrophication process, and to assess the potential for using biological control to improve water quality and increase the yield of harvestable production, the Royal Norwegian Council for Scientific and Industrial Research initiated a research programme on the eutrophication of inland waters. The Eutrophication Programme lasted for 10 years (from 1978 to 1988), involving more than 30 different researchers, resulting in more than 100 scientific papers, and 9 doctoral dissertations.

I had the good fortune of being engaged in the Eutrophication Programme for 2 years (from 1986 to 1988), together with some of the most creative and productive scientists in a new generation of European limnologists. The principal ideas that were conceived in this period are probably still among the major research topics of most of the scientists that were involved. The Eutrophicaton Programme was a scene of active cross-fertilization between the population-oriented approach of classical zooplankton ecology and the system-oriented approach of classical limnology. One of the main concepts resulting from merging these two fields was that the biomass of herbivores, as well as higher trophic levels, must be recognized as significant components of the pelagic nutrient budget. This means that the nutrient household of lakes cannot be considered independently of the trophic structure of the pelagic community, and also that pelagic foodweb structure is closely connected with nutrient supply.

Another key idea developed under the Eutrophication Programme was that nutrient regeneration is an elementary mass balance governed by the differences in element compositions between the grazer and the food it consumes. Although this mass balance approach was introduced and developed in several practically oriented publications (Olsen and Østgaard 1985; Olsen et al. 1986b; Andersen and Hessen 1991), it was felt that there was also an important theoretical potential of this simple and powerful concept. This book has grown out of the difficulties encountered with trying to fully release this potential within the space constraints of scientific journals.

While this book is written in terms of the antiseptic differential equations of theoretical ecology, it is still hoped that it may strike a balance so that it may serve a dual purpose of introducing some of the powerful techniques of mathematical biology to the field ecologist, and some of the contemporary problems in aquatic ecology to the theorist. Although simple models are emphasized throughout this work, some readers might still find the notation to be quite complex at times. As a help and reference to the reader, every chapter therefore contains a list of all symbols used in the presentation, together with definitions and relevant units. Some of the arguments used in this work are based on concepts that might be unfamiliar to the less mathematically inclined. In order to make the presentation more accessible to the general reader, some of the more technical parts have been placed in separate appendices that can be omitted in a casual reading, hopefully without the main ideas being lost.

Many of the ideas presented here resulted from collaborations with other scientists involved in the Eutrophication Programme. I am especially grateful for the continual inspiration and encouragement given by my good friends and coworkers, Dag Hessen, Anne Lyche, Yngvar Olsen and Olav Vadstein. I would also like to thank Arne Jensen, Eystein Paasche and Helge Reinertsen for the ways in which they must have influenced the course of events at times most critical to the completion of this work. Dag Hessen, Anne Lyche, Yngvar Olsen, Eystein Paasche and Tron Frede Thingstad, as well as the series editor, Ulrich Sommer, kindly provided constructive criticism and helpful suggestions on earlier drafts.

The writing of this book was supported by scholarships from the University of Oslo, the Royal Norwegian Council for Scientific and Industrial Research and the Norwegian Research Council. I dedicate it to my wife, Karin, and our children, Katrine and Tone, who patiently endured the intrusion into domestic life represented by the writing of this book.

Oslo, January 1997   Tom Andersen

Contents

Contents

XI

A7 Local Stability Analysis of the Nutrients, Algae,
and Herbivores Model 250
A8 The Persistence Boundary. 258
A9 Numerical Considerations and Computational
Procedures. 261
A10 A Literature Survey of Model Parameters. 265

Subject index 275

1 The Eutrophication Problem in Temperate Lakes: Practical Aspects and Theoretical Ramifications

Scientists are perennially aware that it is best not thrust theory until it is confirmed by evidence. It is equally true ... that it is best not to put too much faith in facts until they have been confirmed by theory.

R. MacArthur

The effects of anthropogenic eutrophication caused by using lakes and rivers as recipients for municipal, agricultural and industrial wastes are often found to be in conflict with commercial and recreational uses of the same localities. If we take Norway as an example, a recent survey of 355 Norwegian lakes (Faafeng et al. 1990) concluded that only 7% of the investigated lakes could be unambiguously classified as eutrophic, of which the majority has probably been eutrophicated by human activity. As this survey was deliberately biased toward lakes located in urban areas, we cannot multiply this figure with the number of lakes in Norway (approximately 400 000 lakes larger than 0.5 ha), to obtain to the total number of eutrophicated lakes. It is not even certain that an unbiased estimate of the number of eutrophicated lakes would be of much value, as the recreational and commercial value to the general public would be expected to be very much higher for lakes located in populated areas. While eutrophication appears to be a problem only on a local scale in some countries, it is still recognized as a major environmental problem in more densely populated areas of central Europe and north America.

Even though wastewater treatment technology is continually improving, the costs of water-treatment processes are usually steeply increasing when the nutrient removal efficiency is approaching 100%. Despite the large investments made by developed countries in wastewater treatment, eutrophication problems will probably nevertheless be something we will have to live with for a long time yet. It might even be argued that complete removal of anthropogenic nutrients is not always a desirable goal; some mild level of nutrient enrichment can, in many instances, be quite beneficial by increasing secondary production, and thereby transforming human wastes to valuable food resources.

For such a view to win acceptance in water resource management, tools must be available for determining what level of nutrient input to a recipient can be tolerated while maintaining a low probability of any negative effects of eutrophication. Today, we do not have enough insight to handle the safe and optimal management of aquatic production systems, and some might argue that this task is so formidable that such a goal will never be reached. On the other hand, the possible gains from having such tools available could be very large if they could be used to differentiate more finely between what would be necessary wastewater treatment in different localities, instead of having a common goal of minimizing nutrient inputs in all recipients. It also should be kept in mind that some major achievements have already been made in our understanding of (at least) freshwater systems since the eutrophication debate started in the early 1960s.

1.1 Eutrophication: Consequences and Correctives

Increasing the input of mineral nutrients to a water body generally leads to an increased standing stock of phytoplankton algae, which again can reduce the water transparency to levels that are unacceptable to many recreational uses. Accompanying the increased nutrient supply, the phytoplankton community is often changed in unfortunate directions with increasing abundance of species producing toxic or obnoxious secondary metabolites that can cause problems for commercial interests like fish farming or drinking water production. Major changes in the food webs often also result from eutrophication, leading to the replacement of valuable fish species with species that are less attractive to commercial or game fisheries.

Both increasing abundance of inedible plankton algae and planktivorous fish will reduce the channeling of nutrients and energy through zooplankton in the planktonic grazer food chains, while more of the primary production will be directed through benthic detritus food chains. Increased benthic metabolism will increase hypolimnetic oxygen consumption and by that increase the chances of internal phosphorus loading from hypolimnetic sediments though redox-controlled phosphorus-binding processes (Mortimer 1941, 1942; Boström et al. 1982).

The reallocation of catabolic processes from the pelagial to the sediments also increases the net CO_2 consumption in the epilimnion. In eutrophicated soft-water lakes the CO_2 consumption from primary production can become so high that it cannot be buffered by the bicarbonate system or compensated for by gas exchange across the air-water interface. Such a breakdown in the bicarbonate buffering system will lead to dramatic pH increase, which again will increase the chances of internal phosphorus loading through pH-dependent desorption of phosphorus bound to littoral sediments (Møller-Andersen 1975; Boström et al. 1982; Jacoby et al. 1982).

One can therefore expect radically different effects on water quality depending on the relative proportions of the phosphorus supply that are accumulated in large, inedible algae or propagated through an efficient grazer food chain.

Phosphorus Loading Models. From both large regional surveys and whole-lake manipulations it has become clear that phosphorus can usually be considered the key limiting nutrient in freshwater plankton systems, and that the phosphorus supply sets an upper bound on the phytoplankton yield of a given lake (Schindler 1977, 1978). Eutrophication control has therefore to a large extent been a question of reducing phosphorus inputs to lakes.

The simple one-box model introduced by Vollenweider (1968) has proven its value as a management tool in predicting the total phosphorus concentration resulting from a given phosphorus loading. The main idea behind this model and its descendants (cf. Reckhow and Chapra 1983) is that the net result of all biotic and abiotic phosphorus transport processes within the lake can be described by a single parameter, *the phosphorus retention*, which is defined as the ratio of the amount of phosphorus held back in the lake to the amount of phosphorus put into the lake. The crucial assumption behind practical application of these models is that the phosphorus retention can be taken to be a static entity for a given lake, and that it can be predicted from easily observable parameters like water renewal time or flushing rate.

By reducing the individual characters of lakes to just a set of morphometric and hydraulic parameters, all aspects of lakes as living ecosystems are neglected. This probably will put strong limitations to what predictive power can ever be gained from this approach, such that further refinement of predictive water quality models will have to include some aspects of biology. On the other hand, the kind of mass-balance philosophy inherent in the simple loading models has served an important purpose in emphasizing lakes as open dynamic systems where the trophic state is a property of the whole watershed and not only of the pelagic zone of the lake (cf. Hutchinson 1973). As such, the dynamic structure of the simple loading models should also provide a sound framework for constructing more complex models where some of the input and output processes are given a biological interpretation.

Water Quality Improvement Through Biomanipulation. Although total phosphorus typically explains more than 80% of the chlorophyll *a* variance in log-log regressions from regional studies, the residual variance in the backtransformed variables will still be sufficient to let the phytoplankton yield vary by an order of magnitude among lakes with the same total phosphorus level. Zooplankton grazing has been identified as a strong modulator of the realized phytoplankton yield in a given locality (Shapiro 1980).

Pace (1984) found that total zooplankton biomass explained less of the residual variance in the relationship between phosphorus and chlorophyll *a* than the relative abundance of large cladoceran grazers, emphasizing the importance of zooplankton community structure on the actual impact of grazing on the phytoplankton community.

Large cladoceran grazers are also preferred prey of most planktivorous fish species, so that their dominance will be possible only at low planktivore abundance (Hrbácek et al. 1961; Brooks and Dodson 1965). This led Shapiro et al. (1975) to introduce the concept of lake restoration by *biomanipulation*, where reduction of the predation pressure on large cladocerans is attempted through either complete removal of fish by, for example, rotenone poisoning, or control of planktivorous fish populations by stocking with piscivorous fish. Successful treatments of this kind have been found to give a plankton community with a high zooplankton biomass dominated by large *Daphnia*, and a reduced phytoplankton biomass dominated by small, fast-growing species that are considered easily edible and assimilable to zooplankton (e.g., Gulati et al. 1990, and references therein).

Since the initial enthusiasm created by the introduction of the biomanipulation concept, several cases have also been reported where food chain manipulations have failed to improve water quality. In some cases with very shallow lakes, the initial problem of excessive phytoplankton growth has simply been replaced by a new problem of excessive growth of benthic algae and macrophytes (e.g., Van Donk et al. 1989). In deeper lakes, failure is usually associated with the invasion of the phytoplankton community by large, inedible species instead of algal species that can be efficiently processed by the planktonic food chains.

Most reported biomanipulation experiments have been performed in isolated lakes of the seepage type, where good phosphorus loading measurements are difficult to obtain. Benndorf (1987) observed that in all biomanipulations that were classified as successful according to his criteria the phosphorus loading was either moderate or unknown, and that biomanipulation should therefore have the greatest likelihood of failure in lakes receiving a phosphorus loading above some critical level. Based on this observation and on experiences from several biomanipulation experiments, Benndorf (1987) suggested that this critical loading might be around 0.2 to 0.4 μg P l^{-1} day^{-1}, although this must be considered as highly tentative until more data from biomanipulations in lakes with different loading conditions have become available.

The immediate effect of an increased grazing pressure will necessarily be reductions in the size of the prey populations of edible algae. Many species have morphological adaptations that reduce ingestibility through a large cell or colony size or digestibility through development of a resistant cell wall or gelatinous sheath, so that different groups of prey species will be captured or utilized with unequal efficiency by grazers. It is therefore not intuitively obvious how a short, efficient food chain from algae to grazers can be made to persist throughout the growing season, and why species with a refuge from grazing losses often seem less favored by high grazing pressure than might be expected. As many species usually classified as inedible to zooplankton, like colony-forming blue green algae, also have very undesirable effects on water quality, the success of biomanipulation in improving water quality must be closely linked to mechanisms that might reduce the competitive ability of inedible algae. It is likely that one key to the suppression of blue green algae in successful biomanipulations lies in the changes in the flows of nutrients accompanying the restructuring of the food web.

1.2 Direct and Indirect Effects of Herbivorous Zooplankton

The Chemostat Analogy of Grazer-Controlled Systems. No animals have 100% efficient utilization of their food, so that grazers will always recycle some unutilized fraction of the nutrients contained in their food and will thus function as both source and sink of phosphorus and other nutrients. From this point of view, an increased grazing pressure will imply an increased turnover of nutrients in the prey compartments in a way that shares some likeness with increasing the flow rate in a chemostat. Decreasing the turnover time or increasing the dilution rate in a population of nutrient-limited algae generally has the effect of decreasing the yield of algal biomass per unit of available nutrient. Reinertsen et al. (1989) found such a reduction of algal phosphorus growth yield driven by increased turnover from grazing to be quite important in improving the water quality after the elimination of fish in Lake Haugatjønna; but from the trophic state concepts of Hutchinson (1967), this would be only an apparent oligotrophication of the system, as the carrying capacity in terms of total phosphorus for potential production of algal biomass would remain the same.

Since the relative abilities of species of plankton algae to succeed in competition for nutrients have been shown to change with both the dilution rate and the supply mode of nutrient (Sommer 1985, 1986; Olsen et al. 1989), changes in the turnover rate resulting from zooplankton grazing also could affect the resource competition between prey species. Selective grazing could either increase or reverse the advantage of the superior competitor for nutrients, depending on whether this species is selected or rejected by grazers (Kilham 1987; Sterner 1989). The chemostat analogy of grazer-controlled systems thus seems to have several interesting aspects that might contribute to explaining why many *Daphnia*-dominated communities apparently are uninvadable to inedible phytoplankton species, but the complex interplay between differential loss rates and competitive abilities cannot be easily analyzed without the help of some modeling effort.

Competition, Stoichiometry, and Resource Supply Ratios. Even with a relatively unselective grazer like *Daphnia* feeding on similarly sized, edible algae, grazing could affect the outcome of resource competition. The resource competition theory of Tilman (1982) predicts that two species of plankton algae can coexist under certain supply ratios of two limiting nutrients, while other supply ratios can lead to the competitive exclusion of one of the species.

Blue-green algae seem to have lower optimal N:P ratios than other phytoplankton algae (Tilman et al. 1982; Smith 1983), so that the supply ratio of nitrogen to phosphorus might have an important effect on the competitive success of blue green algae. From measured elemental ratios in crustacean zooplankton species, Andersen and Hessen (1991) proposed that one should expect a general tendency for higher N:P ratio in recycled nutrients from zooplankton communities dominated by *Daphnia* than from copepod-dominated plankton, so that, in addition to the effects of selective grazing, the zooplankton community structure could have an indirect selection force on the species composition of the phytoplankton community through the N:P ratio of the recycled nutrients.

Resource Partitioning in Food Webs. While much emphasis has been placed on the relationship between algae and phosphorus in water quality management, all living organisms need phosphorus to build vital biomolecules like nucleic acids and membrane lipids. Recent results indicate that the heterotrophic components of pelagic community (bacteria, zooplankton, and fish) have in fact higher requirements for phosphorus than typical plankton algae (Vadstein et al. 1988; Brabrand et al. 1990; Andersen and Hessen 1991).

For example, from the data on zooplankton phosphorus contents in Andersen and Hessen (1991), we can calculate that in situations of heavy *Daphnia* grazing like the spring clear-water phase in Lake Constance (Lampert et al. 1986), up to 40% of total phosphorus can be located in *Daphnia* biomass. Mazumder et al. (1988, 1989, 1990) found that planktivorous fish had a strong influence on the size distribution of particulate phosphorus in both enclosure experiments and natural plankton communities, with more phosphorus being located in zooplankton fractions in the absence of vertebrate predation. This skewing of the phosphorus size distribution toward large organisms like zooplankton was also found to increase the sedimentation of phosphorus (Mazumder et al. 1989) and decrease the hypolimnetic oxygen consumption (Mazumder et al. 1990).

These two observations are consistent if we take into account the comparatively high P:C ratio found in crustacean zooplankton (Andersen and Hessen 1991), although reduced decomposition in the water column of large, fast-sinking organisms, as suggested by Mazumder et al. (1990), might also play a role in reducing the O_2 consumption in the water column per unit of sedimented phosphorus. The results of Mazumder et al. (1988,

1989, 1990) indicate that any extension to the simple loading models, for the purpose of describing biological effects on the phosphorus retention process, should incorporate at least the partitioning of phosphorus between the major biological compartments of the plankton.

Biological Phosphorus Retention Processes. The results of Mazumder et al. (1988, 1989, 1990) fit well with the significant reductions in total phosphorus observed after biomanipulation in, for example, Round Lake (Shapiro and Wright 1984) and Haugatjønna (Reinertsen et al. 1989). If the phosphorus loading stays the same after the biomanipulation, this must be the result of an increased phosphorus retention, or a real oligotrophication of the system sensu Hutchinson (1973). As these lakes were dominated by cladoceran zooplankton, the increased loss of phosphorus was not likely to be caused by "the fecal pellet express" associated with marine copepods (Turner and Ferrante 1979). Wright and Shapiro (1984) suggested that excretion from vertically migrating *Daphnia* could constitute a net transport of phosphorus into the hypolimnion, but both the short gut passage time in *Daphnia* and the common observation that vertical migration ceases when vertebrate predation is relaxed, speak against this explanation.

Although the elimination of fish through poisoning or fishing might remove a substantial fraction of the standing stock of phosphorus in a lake, a change in the size of this pool would normally be undetectable in phosphorus budgets based on total phosphorus estimates from water samples alone. If the fish biomass stands in some kind of long-term equilibrium with the phosphorus loading and if the loading is unchanged after an episodic elimination of fish, the part of the phosphorus load supporting the fish populations probably will be directed elsewhere in the food web. It is only if fish populations are continually harvested that fishing will constitute a retention process in the long-term phosphorus dynamics of the lake.

One of the major effects of reduced vertebrate predation would be a change in the ultimate fate of large *Daphnia* from a high probability of being eaten by fish, to a likelihood of dying from senescence or starvation. From the knowledge of growth efficiencies of planktivorous fish, it would be expected that a large amount of the phosphorus contained in ingested zooplankton would be recycled in the epilimnion (Olsen 1988). On the other hand, zooplankters dying from nonpredatory mortality would sink quickly out from the epilimnion with a velocity of 2 to 20 m h^{-1}, according to data reviewed by Hutchinson (1967), making any decomposition before reaching the sediment surface very unlikely. This was observed in Haugatjønna as a large increase in dead zooplankton collected by the sedimentation traps after the rotenone treatment (Reinertsen et al. 1989).

1.3 Stability and Persistence of Grazer–Controlled Systems

Community Structure and the HSS Hypothesis. While the stability of grazer-controlled plankton communities has direct practical consequences in water resource management, such questions also have a deep relationship to several central problems in ecological theory. The main ideas behind the biomanipulation concept can be traced back to a theory by Hairston et al. (1960), often referred to as the HSS hypothesis, proposing that populations in simple linear food chains appear to be alternately controlled by competition or predation. In food chains with an even number of trophic levels, herbivores would deplete plants and produce barren habitats, while an odd number of trophic levels would release plants from herbivory and produce green habitats where plants become limited by nutrients or other resources.

Oksanen et al. (1981) expanded on this view by a conceptualization where the number of trophic levels is dependent on the richness of the system, and where enrichment sufficient to introduce an additional top predator level would imply a shift from predatory to competitive control, and vice versa, in all the supporting trophic levels. In an alternate model of trophic control, Menge and Sutherland (1976, 1987) argued that the prevalence of omnivory in food webs should lead to increasing control by predation, and decreasing control by resource limitation, for populations at lower trophic levels. The removal of a completely omnivorous top predator would, in the model of Menge and Sutherland (1976, 1987), not be expected to have the same dramatic effects on lower trophic levels as predicted by the HSS model.

McQueen et al. (1986, 1989) challenged the HSS model from a more empirical point of view by correlation analysis of regional and experimental data on total phosphorus and biomasses of algae, zooplankton and fish. Borrowing from the theory of structured programming (e.g., Dijkstra 1976), they coined the phrases bottom-up and top-down control, denoting populations that are controlled by either nutrient supply or predation. From the observed pattern in regression slopes, they inferred a tendency for gradually increasing top-down control and decreasing bottom-up control with increasing trophic level, instead of strict alternations between competitive and predatory control on adjacent trophic levels.

The results of McQueen et al. (1986) initiated a heated debate for some years with sharp frontiers between those supporting fish or phosphorus as the major determinants of phytoplankton biomass in lakes. This argument seems to be settling down now with most participants admitting that there cannot be a single unequivocal answer to such a question, and acknowledging that in a network of interacting components, "effects" may propagate in one direction or another, but not all control comes from either the top or the bottom. Sterner (1989) summarizes the "fish or phosphorus" debate by stating that although grazers do have significant influence on algal community structure and successional trajectories, this does not mean that phytoplankton communities are necessarily and consistently controlled from the top down. The dual role of zooplankton as both sink and source of nutrients through the processes of consumption and recycling implies that the influence of zooplankton on algal communities often comes indirectly through bottom-up factors, i.e., through changed algal resource relations.

Cycles, Bifurcations, and the Paradox of Enrichment. The existence of a critical phosphorus loading above which biomanipulation is unlikely to result in improved water quality, as proposed by Benndorf (1987), seems closely related to a classical result in theoretical ecology called the paradox of enrichment. Rosenzweig (1971) showed that enrichment in the form of increased carrying capacity of the prey population in a simple prey-predator system could destabilize an exploitative equilibrium into a limit cycle that could lead to predator extinction.

Such bifurcation phenomena, or qualitative changes in the behavior of dynamical systems controlled by one or more external parameters, have been the focus of much research in nonlinear dynamics. From more recent work it is known that very complex dynamics can result from very simple systems with three or more state variables (Rogers 1981). For example, Gilpin (1979) showed that in a simple predator-prey model with one predator and two prey species, increasing prey growth rates could drive the system into a characteristic bifurcation sequence where a limit cycle goes through a period-doubling cascade, eventually leading to completely acyclic or chaotic behavior. Similar phenomena have been observed in other ecological models of both predatory (Hogeweg and Hesper 1978) and competitive nature (Arneodo et al. 1982).

While some authors have argued that chaos is common in ecological systems and that this should even lead to a complete revision of ecological theory (Schaffer and Kot 1986), others have claimed that the parameter values necessary to produce chaos are usually far beyond biologically realistic ranges and therefore of minor interest to practical ecology (Berryman and Millstein 1989). Without taking any definitive standpoint in this debate, it is nevertheless clear that the inherent nonlinearity of ecological systems provides a potentially rich repertoire of dynamic behavior, and that anyone attempting to model such systems should at least be aware of the possible complexities that can result from the existence of bifurcation phenomena, multiple steady states, and attractors that are more complicated than simple equilibrium points.

1.4   Scope and Strategy

Many of the central issues in lake management and eutrophication control appear to be deeply related to key research fields in ecological theory, such as resource competition, food web structure, and nutrient cycling. Likewise, the problem of controlling algal blooms by food web manipulations is essentially a question of maintaining the stability and persistence of an enriched prey-predator system. Nevertheless, the exchange of ideas between the fields of lake management and ecological theory appears to have been limited.

This lack of communication might be related to the traditional dichotomy that has separated ecologists into those following the population-community approach and those that follow the process-functional approach (sensu DeAngelis 1990). The first approach, which might be called evolutionary, or species-oriented, focuses on the dynamics of population interactions and the patterns of trophic connections. The second approach, which also can be called flow-oriented or biogeochemical, focuses on the energy and material cycles of ecological systems. While researchers in theoretical ecology have typically been trained in the species-oriented school, researchers working with eutrophication problems have typically been recruited from the flow-oriented school.

Although these two points of view seem to have coexisted in the origins of ecological theory (e.g., Lotka 1925), it is only recently that attempts have been made to reunite them. DeAngelis (1990) points out that food web dynamics and biogeochemistry are deeply interrelated in the sense that while food webs are influenced by limitations of energy and matter, the interactions among species populations within food webs may themselves influence energy flow and material cycling. The present work is an attempt to follow the direction staked out by DeAngelis (1990), by merging some well-established ideas from static nutrient loading models with a dynamic view of pelagic nutrient cycling, and analyzing the resulting models using techniques and principles from the ecological theories of predation and competition.

Time and space constraints exclude treating all aspects of pelagic nutrient cycling in a work like this. By limiting the scope to lakes where planktivorous fish are expected to have minor impact on zooplankton dynamics, a focus on the first two levels of the food web can be defended. This means that the models will primarily apply to lakes where fish are absent either due to food web manipulations or for biogeographic reasons, or where planktivore populations are under strong control by piscivorous fish.

The model development will take a minimalistic approach, implying that simplicity will sometimes be chosen at the price of generality and biological realism. Model analysis will emphasize qualitative dynamical concepts like stability, periodic orbits, persistence, coexistence, and exclusion. Model predictions will be compared with synoptic data on variability among different species or between different lakes, while less weight will be placed on detailed simulation of specific seasonal successions or laboratory experiments.

The model development starts out by reviewing some basic ideas on lakes as open dynamical systems, and their relationship to classical input-output models (Chap. 2). The key concepts from classical loading models are used as the fundament for the formulation of a more general modeling framework, based on stoichiometric relationships in food webs. In the general framework, flows among plankton compartments are presented without reference to the control mechanisms in the pelagic flow network. Two separate chapters (Chap. 3 and 4) are devoted to describing processes regulating population growth and nutrient utilization in specific compartments of the plankton community.

Since primary production constitutes the major autochtoneous energy input to plankton communities, phytoplankton form the base of the food web. Phytoplankton also play a major role in the pelagic nutrient cycles both as exploitative competitors for dissolved nutrients, and as vehicles for transporting nutrients into higher trophic levels, or out of the pelagic zone. Chapter 3 presents a simple model incorporating the necessary interrelationships between phytoplankton growth, nutrient uptake, and nutrient utilization. Emphasis is placed on the observed parameter ranges, and how this variability may be used to construct abstract model species with different competitive traits, while no attempts are made to model specific data sets from any particular phytoplankton species.

Zooplankton communities in lakes with low vertebrate predation pressure appear to have a general similarity in the sense that they are usually dominated by large cladoceran grazers. The simple community structure makes it possible to narrow down the scope of model development to a small set of target species, with members of the genus *Daphnia* as typical representatives. Chapter 4 presents a set of submodels covering different aspects of *Daphnia* biology, on both the individual and the population level. Individual growth is described in relation to both food abundance and nutritional quality, while population growth is described as a function of individual reproductive output and survival probability. The asymptotic properties of the stable age distribution are utilized to construct a simplified model of zooplankton population dynamics without age stucture.

In Chapter 5, modules from Chapters 3 and 4 are assembled into the general framework of Chapter 2, resulting in a minimal model of pelagic nutrient cycling with only three compartments: phosphorus, algae, and grazers. With this particularly simple model structure, it is possible to explore the conditions for local stability, extinction, and persistence by classical analytical and graphical techniques. The analytical results are supplemented by simulations investigating nonlinear phenomena like bifurcations and periodic orbits. The model predictions are used to construct loading criteria and trophic state indicators for biomanipulated lakes.

Chapter 6 presents several examples of how the minimal model of pelagic nutrient cycling can be extended stepwise toward a closer resemblance to a real pelagic food web. The base of the food web is widened by introducing new algal species with different competitive abilities and different edibilities to the zooplankton. The interaction of two different nutrient cycles is studied by introducing nitrogen as a second, potentially limiting nutrient. Carbon cycling is represented by introducing detritus as an alternate food source to the zooplankton. The interaction of carbon and nutrient cycling is studied by introducing bacteria as both competitors for mineral nutrients and as salvagers of detritus carbon. The effect of each extension is compared to the minimal model in Chapter 5, with special emphasis on the possible emergence of new modes of dynamic behavior as the number of state variables is increased.

A short concluding chapter then tries to summarize the major results of the preceding ones, and also discuss some of the limitations and omissions in the models. In order to test model predictions and improve on model limitations, several directions for future research are suggested.

2 The Biogeochemical Theatre – Phosphorus Cycling and Phosphorus Household in Lakes

> *For the drama of life is like a puppet show in which stage, scenery, actors and all are made of the same stuff. ... and if we would catch the spirit of the piece, our attention must not be absorbed in the characters alone, but must be extended also to the scene, of which they are born, on which they play their part, and with which, in a little while, they merge again.*
>
> Alfred James Lotka (1925)

Except for certain arid regions where lakes lose water only by evaporation, lakes are open flow-through systems whose content of a given element is determined by the dynamic balance between supply through the inflows, loss through the outflows, and internal sources and sinks. At sufficiently long timescales, the lake content of a given element tends to an equilibrium depending on the properties of both the lake and its catchment area. For dissolved substances that are unaffected by biological transformations (conservative elements), inflows and outflows can be different only in transient situations such as after a change in input concentration.

For mineral nutrients like N and P, net uptake by phytoplankton and subsequent transfer through the food web will deplete the concentration of dissolved chemical species in the recipient water, so that dissolved nutrient concentration in the inflows is always likely to exceed that in the water body. The biological transformation from a dissolved to a particulate state also makes the fate of nutrient atoms to be influenced by the force of gravity, thus increasing the probability of an atom being lost from the water column and permanently buried in the sediments. The retention of mineral nutrients in lakes has important effects on the dose-response relationship between nutrient loading and lake nutrient content, and is therefore considered a key element in the kind of nutrient mass-balance models that are commonly used in water quality management.

2.1 Phosphorus Mass–Balance Models

Based on the broad consensus that phosphorus is the mineral nutrient most likely to become limiting to the yield of primary producers in lakes (Schindler 1977, 1978), controlling P inputs and predicting responses to changes in P loading have been among the major concerns of lake management. Prediction is commonly based on a family of phosphorus mass-balance models directly descending from the pioneering work of Vollenweider (1968).

The total lake content of phosphorus is the product of the lake volume (V; l) and the total P concentration in the lake [P_T; (μg P) l^{-1}], while the total P loading rate is the product of the volume-averaged concentration in the inflowing water [P_L; (μg P) l^{-1}] and the water flow into the lake (q_i; l day^{-1}). If we adopt the assumption that concentrations in the lake and in the outflowing water are equal, which is often called the continuously stirred tank reactor (CSTR) assumption in engineering literature (e.g., Reckhow and Chapra 1983), phosphorus will be lost through the outflow at a rate which will be proportional to the water flow rate out of the lake (q_o; l day^{-1}) and the concentration in the lake (P_T). For nonconservative elements like phosphorus, there will also be net losses or gains due to internal processes in the lake [S_P; (μg P) l^{-1} day^{-1}]. The rate of change in lake content will be the difference between gains from the inflow, losses through the outflow, and the contribution from internal sources and sinks (here, and elsewhere in the forthcoming sections, the notation $\dot{x}$ will mean the derivative of x with respect to time):

$$\left(\dot{VP_T}\right) = \dot{V} P_T + V \dot{P_T} = q_i P_L - q_o P_T - V S_P. \tag{2.1}$$

The first identity in Eq. (2.1), which is simply the derivative of a product, takes care of the fact that a change in either lake volume or lake concentration will lead to a change in lake content. The rate of change in lake volume is equal to the difference between inflow and outflow ($\dot{V} = q_i - q_o$), such that the maintenance of a constant lake volume would require that $q_i = q_o = q$. Under the assumption of equal in- and outflows, the mass-balance equation (2.1) can be simplified as

$$\dot{P_T} = D\left(P_L - P_T\right) - S_P, \tag{2.2}$$

where D is the hydraulic loading rate ($D = q/V$; day^{-1}), or the dilution rate of the system. If we define the volumetric phosphorus loading rate as $L_p = D P_L$ [in units (μg P) l^{-1} day^{-1}], the steady-state solution to Eq. (2.2) can be written as

$$P_T = \left(1 - \frac{S_P}{L_P}\right) P_L = \left(1 - R_P\right) P_L \tag{2.3}$$

where the dimensionless quantity R_p is termed the phosphorus retention of the lake, or the ratio of P retained in the lake (S_p) to P entering the lake (L_p). Equation (2.3) implies that predicting the response in terms of lake total P concentration from a given dose in terms of P loading is equivalent to predicting the phosphorus retention of the lake. It is, therefore, not surprising that much research effort has been focused on predicting R_p for a given lake from other, more easily measurable morphometric and hydraulic parameters.

The annual average water renewal rate of a temperate lake will necessarily conceal substantial temporal variation; since much of the annual water budget will be associated with the spring flood, throughflow will usually be much less in other seasons. An annual average dilution rate of 0.01 day^{-1} would thus correspond to a much lower dilution rate experienced by the summer plankton community. If we are mostly concerned with water exchange rates in the stratified season, then we must also take into account water exchange between epi- and hypolimnion due to deepening of the mixed layer (entrainment). If mixed layer depth doubles from start to end of the stratified season (5-6 months in northern temperate areas), this will be equivalent to an entraiment-driven dilution rate <0.0066 day^{-1}. In other words, dilution loss rates due to throughflow and entrainment are most likely in the order of <1% day^{-1} for plankton communities of stratified, temperate lakes. As will be apparent from forthcoming chapters, such a rate will be small compared to maximal specific rates for most biological process in plankton communities.

2.2 Phosphorus Retention in Lakes

Apart from certain transient situations occurring after strong reductions of P loading to highly eutrophic lakes, the phosphorus retention will generally be positive. This means that over sufficiently long timescales, all lakes will act as net sinks for phosphorus. Mass-balance studies in many lakes show that phosphorus retentions over the whole domain $0 \le R_p \le 1$ can be measured, making the predictive power of Eq. (2.3) essentially zero as long as R_p is unknown for a particular lake. Vollenweider (1976) observed that phosphorus retention varied systematically with water residence time, so that an increase in the water residence time would increase the probability of a given P atom being trapped in the bottom sediments instead of leaving the lake via the outflow. Several authors have since presented slightly different functions fitted to empirical data sets for the purpose of predicting P retention when water residence time or dilution rate is known (e.g., Oglesby 1977; Lee et al. 1978).

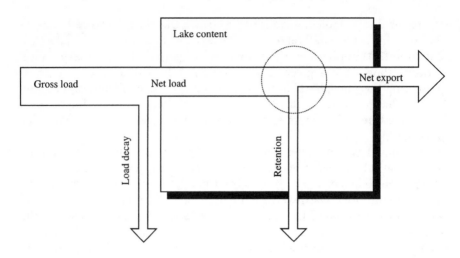

Fig. 2.1. Flow diagram for a typical phosphorus loading model including the the load decay concept of Prairie (1988). *Dotted circle* encloses the domain of the internal phosphorus cycle in the lakes, as elaborated in Fig. 2.6

A particularly simple and elegant interpretation of the variability in phosphorus retention among lakes was presented by Prairie (1988), who proposed that the P retention could be partitioned into two components; losses directly from the P load and losses from the lake P content (Fig. 2.1). The load decay part is thought to result from rapid sedimentation of coarse mineral particles and perhaps also rapid uptake by macrophytes as the inflowing water enters the lake. This concept is supported by, among others, the results of Emondson and Lehman (1981), who found the difference between P load and P outflow in Lake Washington to be much larger than the actual measured sedimentation, indicating that a fraction of the P load was lost rapidly upon entering the lake and therefore not captured by the sediment traps. If we denote the fraction of the load that is lost before entering the lake by R_L and the net sinking loss rate of total P in the lake by σ_P (day^{-1}), the internal P loss rate can be written as

$$S_P = R_L L_P + \sigma_P P_T \qquad (2.4)$$

Substituting Eq. (2.4) into Eq. (2.3) gives a pair of equations which can be solved with respect to S_P, yielding the rational function

$$R_P = \frac{\sigma_P + R_L D}{\sigma_P + D}. \qquad (2.5)$$

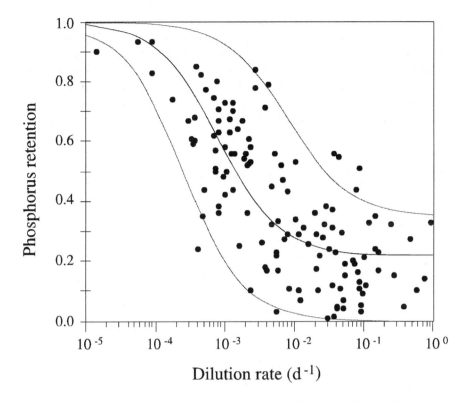

Fig. 2.2. Observed and predicted phosphorus retention in lakes with different dilution rates (data set compiled by Prairie 1988; n = 120). *Solid line* Fitted average relationship according to Eq. (2.5); *broken lines* lower and upper limits of approximate 95% confidence band for individual observations

In lakes with very long water residence times, nearly all of the P input will be lost to the sediments ($R_P \to 1$ as $D \to 0$), while in lakes with very high flushing rates there will be almost no sedimentation loss; only load decay through settling of coarse particles and uptake by rooted vegetation will affect the P retention ($R_P \to R_L$ as $D \to \infty$).

In an analysis of phosphorus budgets from 120 different lakes, with dilution rates spanning five orders of magnitude, Prairie (1989) fitted Eq. (2.5) to measured P retention, with parameter estimates $R_L = 0.22$ and $\sigma_p = 0.0008$ day^{-1}. In other words, for the average lake, 22% of the P load is lost before entering the lake, while only 0.08% of lake total P is lost to the sediments per day (or 29% per year). Although Eq. (2.5) certainly captures the trend in the data set shown in Fig. 2.2, there is still a large residual variance. For example, at a typical dilution rate of 0.01 day^{-1} observed P retention varies from < 0.05 to > 0.5, or more than an order of magnitude. For a pair of curves described by Eq. (2.5), the parameters can be adjusted so that the two curves symmetrically enclose the central 95% of the observations in Fig. 2.2; thus constituting a crude 95% confidence band for the individual observations in the data set. The envelope curves shown in Fig. 2.2 correspond to a variation in R_L from 0 to 0.35, and in σ_P from 0.00025 day^{-1} to 0.008 day^{-1}. In other words, the load decay can be negligible in some lakes while constituting more than one third of the load in others, while the net sedimentation loss rate σ_P can vary by a factor of 32 among lakes with similar hydrological properties.

2.3 Biologically Mediated Phosphorus Retention Processes

Phosphorus losses from the water phase to the sediments will usually be closely linked to the settling of suspended particles, although exceptions might be found in shallow lakes with a large ratio of sediment surface area to lake volume, where direct adsorption/desorption processes between water and sediment might be of importance. In lakes with nonreducing hypolimnia, the P flux associated with settling particles will largely be a net loss process, as phosphorus will normally be tightly bound to inorganic sediment particles or incorporated in the biota of the sediment. While inorganic particles like calcite and iron hydroxides or allochtoneous organic particles like humus aggregations can be an important part of the sedimenting material flux in certain types of lakes, it seems reasonable to assume that the majority of the settling particles will be organic and of autochtoneous origin. This means that much of the variability in phosphorus loss rates among lakes with similar hydrology should be related to the partitioning of phosphorus between particles with different residence times in the pelagic zone.

The steady-state loss rate of passively sinking particles from a homogeneous pelagial is given by $\sigma = u/z_m$, where u is the terminal sinking velocity (m day^{-1}) and z_m is the depth of the mixed layer (m). The terminal sinking velocity of a spherical particle is determined by the balance between viscous and gravitational forces, which can be expressed as a function of particle diameter and density by Stokes law. For nonspherical particles the Stokes equation needs to be adjusted by a form-resistance factor, accounting for the fact that, for example, elongated, needle-like particles have much lower sinking velocities than spherical particles of equivalent volume and density (Hutchinson 1967).

In addition to the morphologically determined form-resistance factor, plankton organisms have several other adaptations to remaining suspended in the water column. Flagellated phytoplankton species, ciliates, and all planktonic metazoans have the ability to counteract the force of gravity by active motion. Several nonflagellated groups of plankton algae have acquired the ability to modulate their sinking velocity through buoyancy regulation. This ability is most highly evolved in colonial cyanobacteria, where the regulation of density by means of gas vesicles enables this group to perform active diurnal migration (Walsby and Reynolds 1980). More limited changes in sinking velocity related to growth conditions have also been found in other, nonflagellated groups like diatoms (Titman and Kilham 1976).

The basic principles of the Stokes relation apply equally to organisms with positive and negative buoyancy, so that both terminal upward migration velocity and terminal sinking velocity will generally increase with cell or colony size (Reynolds 1989). It is, therefore, only the largest phytoplankton units like colonial cyanobacteria and large dinoflagellates that are able to perform significant diurnal vertical migration, enabling them to exploit hypolimnetic nutrients resources that would be unavailable to nonmigrating species. It is thus conceivable that in certain lakes, vertically migrating species of algae can constitute a net source of phosphorus to the pelagic zone.

Vertical migration is also well documented in the major groups of macrozooplankton. Although a variety of theories have been presented to explain this behavior (see, e.g., Mangel and Clark 1988), the current consensus seems to be that vertical migration is mainly the result of a tradeoff between foraging and predator avoidance (Gliwicz 1986; Gliwicz and Pijanowska 1989). Wright and Shapiro (1984) proposed that the reduction in total phosphorus after biomanipulation in Round Lake was caused by vertically migrating *Daphnia* that were excreting phosphorus in the hypolimnion at night, which was ingested in the epilimnion during the day. From the close coupling between ingestion and release of phosphorus that has been found in *Daphnia* (Olsen and Østgaard 1985; Olsen et al. 1986b) and the short gut passage times found in *Daphnia* (Geller 1975), it is unlikely that *Daphnia* P release in the hypolimnion should be related to P ingestion in the epilimnion for more that a short transient period (a few gut evacuation periods) after migrating. Vertically migrating zooplankton should therefore have only minor impact on the phosphorus budget of the pelagic zone.

In a review of measured sinking rates in plankton algae, Heaney and Butterwick (1985) indicate sinking velocities $\sigma = 0.1 - 1$ m day^{-1} in large, nonmotile plankton algae like diatoms and desmids. Such sinking velocities would translate to loss rates from 0.01 to 0.2 day^{-1} for organisms suspended in a typical mixed layer of depth $z_m = 5\text{-}10$ m, which is more than an order of magnitude larger than the average sinking loss rate as estimated by Prairie (1989) and also above the 95% confidence band in Fig. 2.2. This indicates that the phosphorus economy of lakes is far more efficient than would be the case if large, nonmotile plankton algae were a major compartment of total phosphorus.

One would expect motile members of the phytoplankton community to suffer negligible sinking losses while alive. If nonpredatory mortality is important in the dynamics of plankton algae, as suggested by some authors (e.g., Jassby and Goldman 1974), dead cells of both motile and nonmotile species of the same size would be expected to be lost from the mixed layer at a similar rate. The effect of predatory mortality on the residence time of algal material in the mixed layer is, among other things, dependent on the characteristics of the grazer.

The fecal material from calanoid copepods and euphausids remains surrounded by a peritrophic membrane after egestion, constituting discrete units called fecal pellets. The grazing activity by marine macrozooplankton is therefore expected to increase the nutrient losses from the plankton community through the repackaging of smaller, both motile and nonmotile, algal cells into larger, faster-sinking fecal pellets (Turner and Ferrante 1979). The peritrophic membranes of *Daphnia* and other freshwater cladocerans are disrupted before egestion so that feces are released as a flocculent substance composed of very small, and therefore slowly sinking, particles (Peters 1987). We can thus expect that in freshwater plankton communities, the presence of cladoceran grazers will tend to increase the average residence time of dead or nonmotile algal material in the mixed layer by shredding and disrupting larger cells and aggregates into smaller particles.

Detritus will often be the major component of the suspended organic matter - even in lakes that receive negligible amounts of allochtoneous particle input. It seems that zooplankton feces, especially from cladoceran grazers, can be a major source of autochtoneous detritus (Olsen et al. 1986a; Hessen et al. 1990). Feces will generally be impoverished in nutrients compared to the food, as grazers are able to extract a major fraction which can either be released as recycled nutrients or incorporated into grazer tissue. Repeated reingestion of detritus will, as discussed by Hessen and Andersen (1992), lead to a progressive reduction of its nutrient content. While grazing indisputably can be a major process in the vertical carbon flux in the oceans (Banse 1990) and in the production and destruction of detritus carbon in lakes (Hessen et al. 1990), these carbon fluxes will not necessarily be mirrored by nutrient fluxes of comparable magnitude – at least not in lakes dominated by cladocerans.

If we interpret the inverse of the maximal sinking loss rate as the minimal residence time in the mixed layer, we can expect dead algal cells and fecal material on the average to remain suspended in the mixed layer for at least 5 days before being lost to the hypolimnion. During this time, suspended detritus can be the victim of autolytic processes, microbial degradation, and reingestion by grazers, all of which would tend to extract nutrients from the detritus before it leaves the mixed layer. Some field evidence for such an efficient reclamation of phosphorus from dead algal cells and fecal material is given by Olsen et al. (1986b), who found by use of X-ray micrography analysis that suspended detritus particles have very low P content.

Compared with nonmotile algal cells, dead zooplankton organisms sink very fast; dead or narcotized *Daphnia* have sinking velocities of 50 to 500 m day^{-1} according to data reviewed by Hutchinson (1967), corresponding to an average residence time of at most 5 hours in the mixed layer after death. It thus seems much more likely that the nutrients in dead algal cells could be salvaged by autolysis and microbial degradation before leaving the mixed layer, than in dead zooplankton material. The fate of zooplankton-bound nutrients depends on the dominant cause of zooplankton mortality. If predatory mortality is high, then nutrients will most likely be retained in the pelagic zone either in the form of predator tissue or as nutrients recycled to the plankton community. If nonpredatory mortality due starvation or senescence is dominating, the majority of zooplankton-bound nutrients will eventually be lost from the plankton community. The potential importance of the latter process is emphasized by the observation that in lakes with low zooplankton predation pressure, zooplankton can constitute a significant part of the vertical flux of phosphorus (Reinertsen et al. 1989; Mazumder et al. 1990).

2.4 Partitioning of Phosphorus in the Plankton Community

In lake management, the prediction of lake total phosphorus from loading and retention is normally just a vehicle for predicting algal biomass, which again is closely related to common water quality criteria like turbidity and transparency. The general tendency for algal biomass (usually measured as chlorophyll *a* concentration) to increase with total phosphorus is well established, although most total P/chla regressions are found to contain considerable residual variance. Figure 2.3 shows that while a 1:1 relationship between chla and total P forms a reasonable upper bound to potential algal yield at given total P level (delimiting 99.1% of the observations), the variance in the actual chla yield is more than one order of magnitude.

Many attempts have been made to reduce the prediction variance of algal biomass in water quality models, most of them based on the assumption that the variability in algal yield at a given total P level is mainly caused by qualitative differences in the phosphorus supplied to different lakes. Much research effort has therefore been invested in the quantification of the biologically available part of the phosphorus supply (cf. reviews by Cembella et al. 1984a,b), although no generally accepted way to define bioavailable P seems to have resulted from this. Much of the resulting confusion might be rooted in the problems of defining a relevant timescale for bioavailability. From the results of short-term bioassays, it has been argued that only the pools of inorganic and easily hydrolyzable/desorbable phosphorus that are

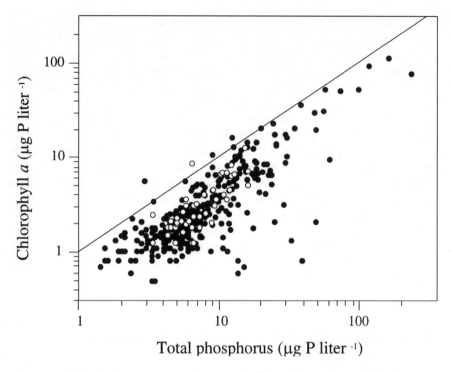

Fig. 2.3. Relationship between seasonal means of chlorophyll *a* and total phosphorus in Norwegian lakes. *Solid circles* Regional survey data from Faafeng et al. (1990); *open circles* data from Hessen et al. (1992); *solid line* 1 : 1 relationship between chlorophyll *a* and total P

measured with the common molybdate-blue method can be considered bioavailable (e.g., Løvstad and Wold 1984). At the other extreme, the load decay estimate of Prairie (1988) indicates that as much as 78% of the P load can be considered potentially bioavailable in the average lake.

Rigler (1964) was among the first to notice the relative constancy in the distribution of phosphorus among commonly analyzed fractions. Fig. 2.4 shows the distribution between major dissolved and particulate P fractions in a subset of 45 of the 355 lakes surveyed by Faafeng et al. (1990; Fig. 2.3). In this study the particulate fraction ranged quite symmetrically from 22 to 76% of total P around a median of 48%. Under the assumption of a constant algal yield per unit particulate P, this allows variation by a factor of < 4 in chlorophyll *a* at a given total P level, or considerably less than the observed variance in Fig. 2.3. The concentrations of molybdate-reactive phosphorus were in all cases close to analytical detection limits in the data set of Hessen et al. (1992). As many studies indicate that the molybdate-blue method tends to overestimate the concentration of orthophosphate, it seems reasonable to assume that dissolved inorganic phosphorus constitutes considerably less than the 10% of the total indicated in Fig. 2.4.

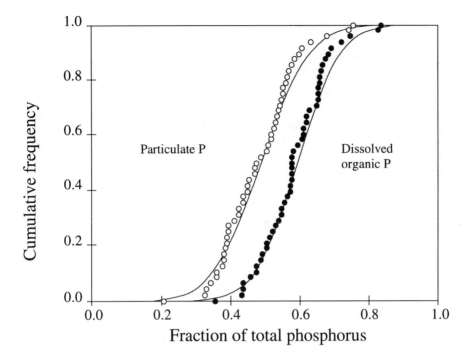

Fig. 2.4. Cumulative frequency distributions of the major fractions of total phosphorus. Seasonal averages from a regional survey covering 45 Norwegian lakes reported by Hessen et al. (1992). *Open symbols* Particulate fraction of total P; *solid symbols* particulate + dissolved inorganic fractions of total P; *solid lines* fitted normal distributions

The major part of the dissolved P in the study of Hessen et al. (1992) was in the form of dissolved organic P, an operationally defined pool consisting of the part of nonmolybdate reactive phosphorus that passes through a GF/C filter. It is likely that this pool includes a significant fraction of both small bacteria and material leaking from delicate cells that break up on the filter (Taylor and Lean 1991). The relatively low variability in the partitioning between the dissolved and particulate fractions (Fig. 2.4) might result if a major part of dissolved organic P is, in fact, particulate P that has been dislocated by methodical artifacts. It could also be explained by the slow exchange between dissolved organic P and other phosphorus pools that has been inferred from tracer studies (Lean 1976). In any case, all the major phosphorus pools seem to stand in some sort of dynamic relationship to each other, so that none of them can be considered entirely refractory.

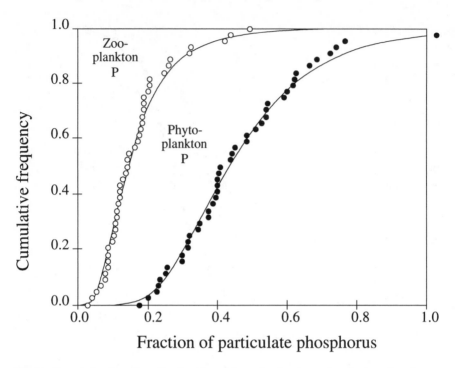

Fig. 2.5. Cumulative frequency distributions of the major fractions of particulate phosphorus. Seasonal averages from a regional survey covering 45 Norwegian lakes reported by Hessen et al. (1992). *Open symbols* Fraction of particulate P in zooplankton; *solid symbols* fraction of particulate P in zooplankton + phytoplankton under the assumption of Redfield P:C proportions in algae; *solid lines* fitted lognormal distributions

It is reasonable to expect that both the phosphorus retention and the yield of algal biomass in a given lake will be closely linked to the partitioning of phosphorus within the >60% of total P constituted by the particulate and labile dissolved fractions. The phosphorus content of planktonic microorganisms is generally found to be related to the growth conditions, as has been repeatedly demonstrated in algae (Droop 1983; Turpin 1988), and also in bacteria (Vadstein et al. 1988) and unicellular microzooplankton (Andersen et al. 1986). Both bacteria and algae are able to utilize P efficiently when the supply is low and to store quantities in excess of their immediate needs when the supply is high (luxury uptake). Bacteria seem to have a P demand almost an order of magnitude higher than algae (Vadstein et al. 1988), while zooplankton takes an intermediate position between these extremes (Andersen and Hessen 1991). Although some means of economizing with phosphorus seem possible also in metazoan zooplankton, the phosphorus content in herbivorous zooplankton seems to be much less variable than in the particles on which they feed (Andersen and Hessen 1991).

Hessen et al. (1992) estimated the zooplankton contribution to particulate phosphorus in 45 Norwegian lakes, based on zooplankton biomass measurements and species-specific conversion factors from dry weight to P. As shown in Fig. 2.5, zooplankton phosphorus constituted from 3 to 49% of particulate P, with a median of 14%. The distribution was more skewed than the distributions of total P fractions (Fig. 2.4), and could be reasonably well represented by a lognormal distribution.

The phytoplankton contribution to particulate P is more difficult to quantify, but a rough maximum estimate is possible from phytoplankton biovolume measurements and the assumption that the algae contain C and P in Redfield proportions (atomic C:P = 106:1). The Redfield ratio is generally thought to be characteristic of algae growing at a rate close to their innate capacity (e.g., Goldman et al. 1979), so that slower-growing algae would be expected to contain less P than this per unit biomass. The resulting sum of zoo- and phytoplankton contributions ranged from 18 to >100% of particulate P, with a median of 41% (Fig. 2.5). The presence of two cases where this maximum estimate exceeded the measured particulate P is not unexpected, when taking into account the accumulated measurement uncertainties involved in this calculation, though it might also be noticed that these two lakes are among those that had the highest yield of the chlorophyll a per unit total P in Fig. 2.3, indicating that algal growth rates might have been appreciably below their potential.

A main feature of Fig. 2.5 is that even when applying a maximum estimate for the algal P:C ratio, the compound contribution from phyto- and zooplankton amounted to <60% of particulate P in 75% of the investigated lakes. Although the remainder may have some contribution from nonliving particles (like clay or humus), it is most likely that a significant fraction of it is located in bacteria (cf. Vadstein et al. 1988). It thus seems reasonable to conclude that, while a significant fraction of particulate P is allocated to phytoplankton, it is often not the major phosphorus pool in the plankton community. On the other hand, both bacteria and zooplankton seem to contain more conspicuous fractions of particulate P than would be expected from their contributions to plankton biomass.

2.5 The Dynamics of the Pelagic Phosphorus Cycle

We can expect different components of the plankton to suffer widely different sinking losses, making the actual sedimentary loss of phosphorus in a given lake crucially dependent on both the partitioning of phosphorus between dissolved and particulate fractions, and among different groups of plankton organisms. The distribution of phosphorus within the particulate fraction will depend on both the biomass proportions among different plankton components and their specific phosphorus contents. The biomass

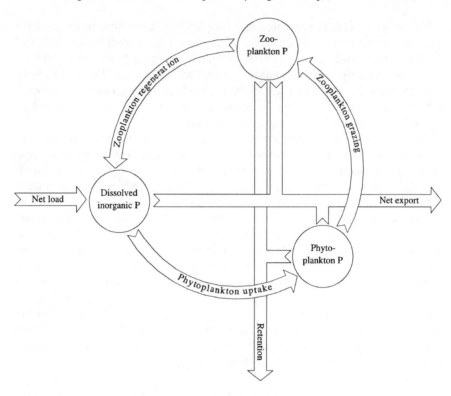

Fig. 2.6. Flow diagram illustrating the major flows of the internal phosphorus cycle, and their relationship with the mass balance of phosphorus in the lake. Location within the lake-wide phosphorus budget is indicated by the *dotted circle* in Fig. 2.1

proportions between plankton components will also affect the realized phytoplankton yield, and thus influence the apparent water quality resulting from a given phosphorus loading.

The partitioning of phosphorus among the biologically important forms will be the net result of a dynamic balance between the main flows of the phosphorus cycle. Figure 2.6 illustrates how the major terms of the phosphorus budget (loading, retention, and export through the outflow) can be linked to main processes of the internal phosphorus cycle (uptake, grazing, and regeneration).

In accordance with the declared scope of this work, the conceptual model in Fig. 2.6 represents only the first two levels of the food web. This will be only a rough approximation of many plankton communities, where the food web can also include invertebrate zooplankton predators, and both planktivorous and piscivorous fish. Although a major fraction of lake total phosphorus can be contained in fish biomass (e.g., Brabrand et al. 1990; Mazumder et al. 1992), the comparatively slow turnover of this pool makes the fish play a less conspicuous role in the phosphorus cycle than would be expected from their share of the standing stock of phosphorus (Nakashima and Leggett 1980; Den Oude and Gulati 1988; Mazumder et al. 1992). While the direct effects of planktivorous fish on the phosphorus cycle are omitted in the present framework, some indirect effects of restructuring the zooplankton community by selective vertebrate predation can still be represented as species substitutions in the zooplankton compartment.

It should be noticed that, while Fig. 2.6 contains no explicit bacterial compartment, the current consensus on the role of bacteria in the pelagic phosphorus cycle can still be represented in the framework of the conceptual model. If bacteria function more as exploitative competitors for nutrients than as nutrient remineralizers (Currie and Kalff 1984; Güde 1985; Thingstad 1987), their role in the phosphorus cycle should be functionally equivalent to the phytoplankton compartment in Fig. 2.6, even if their role in the carbon cycle is very different from phytoplankton. The zooplankton compartment is sufficiently general that also specialized bacterial predators like heterotrophic flagellates can be represented. On the other hand, the simple flow scheme in Fig. 2.6 is unable to represent omnivory and mixotrophy in the phosphorus cycle, despite the inferred importance of organisms with such feeding modes in some recent studies (Bird and Kalff 1986; Porter 1988).

Even in the simple system depicted in Fig. 2.6, the close couplings between the intercompartmental flows make it very difficult to make any predictions on the partitioning of phosphorus without some kind of dynamic model of the phosphorus cycle. If we assume that the plankton community is composed of a set of algal populations with biomasses $C_1, C_2, ...$ [(mg C) l^{-1}] and a set of zooplankton populations with biomasses $Z_1, Z_2, ...$ [(mg C) l^{-1}], total P [P_T; (µg P) l^{-1}] can be written as the sum of inorganic P [P_I; (µg P) l^{-1}] and P contained in all algal and zooplankton populations

$$P_T = P_I + \sum_i Q_i C_i + \sum_j \theta_j Z_j , \qquad (2.6)$$

where Q and θ_j are the phosphorus contents [(µg P) (mg C)$^{-1}$] of phytoplankton population i and zooplankton population j, respectively. Physiological studies on plankton algae generally show that algal P content is a variable quantity related to growth rate and possibly other factors. The decoupling of algal P uptake and C fixation makes it necessary to include two state variables in order to describe the P and C dynamics of a phytoplankton population. If we choose the biomass of population i (C_i) as the first state variable, and the concentration of P contained in population i [P_i; (µg P) l^{-1}] as the second state variable, the P and C dynamics of population i can be written as

$$\dot{P}_i = v_i C_i - D_i P_i , \qquad (2.7)$$

$$\dot{C}_i = (\mu_i - D_i)C_i. \tag{2.8}$$

The differential equation (2.7) states that the net rate of change in P contained in population i is the difference between uptake and loss, where V_i is the net, biomass-specific P uptake rate [(μg P) (mg C)$^{-1}$ day^{-1}] and D_i is the aggregated loss rate (day^{-1}) from population i. Likewise, Eq. (2.8) says that the net rate of change in the biomass of population i is the difference between growth and loss, with μ_i being the net, specific growth rate (day^{-1}) of the population. The aggregated loss rate D_i is the sum of population losses due to dilution, sinking, and grazing:

$$D_i = D + \sigma_i + \sum_j F_{ij} Z_j. \tag{2.9}$$

The second term in Eq. (2.9) is the sedimentation loss rate of population i (σ_i; day^{-1}), while the last term is the summation of grazing losses from population i over all zooplankton populations. F_{ij} is the specific clearance rate [1 (mg C)$^{-1}$ day^{-1}] of zooplankton population j when feeding on phytoplankton population i. The clearance rate can be interpreted as the virtual volume of water from which all food particles can be removed by the activity of a unit grazer biomass in a unit of time. The equality of the loss terms in Eqs. (2.7) and (2.8) is equivalent to the reasonable assumption that phytoplankton cells are lost as entire units, irrespective of whether the loss is effected by dilution, sinking, or grazing.

The P and C dynamics of an algal population [Eqs. (2.7) and (2.8)] can also be expressed with algal biomass (C_i) and P content (Q_i) as state variables, which is more common in algal chemostat models (e.g., Olsen et al. 1989). From the relationship $P_i = Q_i C_i$, the time derivative of Q_i can be found from the rule for taking the derivative of a ratio:

$$\dot{Q}_i = \left(\frac{\dot{P}_i}{C_i}\right) = \left(\frac{\dot{P}_i}{P_i} - \frac{\dot{C}_i}{C_i}\right)Q_i. \tag{2.10}$$

Substitution of the expressions for $\dot{P}_i$ and $\dot{C}_i$ [Eqs. (2.7) and (2.8)] into Eq. (2.10) gives

$$\dot{Q}_i = v_i - \mu_i Q_i, \tag{2.11}$$

which is identical to the equation developed by Droop (1974), stating that the rate of change in P content is determined by the balance between uptake and growth.

From the discussion in Section 2.4, it is reasonable to assume that the P content of zooplankton is a fixed, species-specific quantity. The implied balance between P and C allocation in zooplankton (which is equivalent to $\dot{\theta}_j = 0$ for all j) means that a single state variable is sufficient to describe both the P and C dynamics of a zooplankton population. If we choose biomass (Z_j) as the state variable, the dynamics of zooplankton population j

will be determined by the balance between growth and loss through the differential equation

$$\dot{Z}_j = [g_j - (\delta_j + D)]Z_j,$$ (2.12)

where g_j and δ_j are the specific growth and mortality rates (day^{-1}) of population j.

Although the pool of dissolved inorganic phosphorus will almost always be a minor fraction of total P, this compartment serves as the key link between regeneration and uptake and will be an essential part of every model describing exploitative competition for phosphorus between phytoplankton species. Without the interference of plankton organisms, the net rate of change in the inorganic P concentration (P_I) will be as for a conservative substance: the difference between the external supply rate and the loss rate through the outflow [the first term in Eq. (2.13)]. The activities of plankton organisms both remove inorganic P through phytoplankton uptake [the second term in Eq. (2.13)] and supply inorganic P through zooplankton regeneration [the third term in Eq. (2.13)]:

$$\dot{P}_I = D\left(P_L - P_I\right) - \sum_i v_i C_i - \sum_i \rho_i Z_i.$$ (2.13)

The phosphorus regeneration rate [ρ_j; (µg P) (mg C)$^{-1}$ day^{-1}] of a given zooplankton population is determined by the difference between P intake through ingested food and P utilization into the production of new zooplankton biomass (Olsen and Østgaard 1985), which can be written as

$$\rho_j = \sum_i F_{ij} P_i - g_j \theta_j.$$ (2.14)

The first term in Eq. (2.14) corresponds to the total food P intake in zooplankton population j, summed over all prey populations, while the second term is the biomass-specific rate of P incorporation into new zooplankton biomass.

Equations (2.7) to (2.9) and (2.12) to (2.14) together describe the phosphorus cycling in a plankton community consisting of an arbitrary number of phyto- and zooplankton populations. To see how the internal cycling in the plankton system relates to the total phosphorus budget of the lake, one can proceed by taking the derivative of Eq. (2.6), giving the following relation between net changes in individual P compartments and net change in total P:

$$\dot{P}_T = \dot{P}_I + \sum_i \dot{P}_i + \sum_i \theta_i Z_i.$$ (2.15)

Substituting the timederivatives on the right hand side of Eq. (2.15) with the relevant mass-balance terms given by Eqs. (2.7), (2.12), and (2.13) leads to a massive cancellation of factors such that Eq. (2.15) can be written as

$$\dot{P_T} = D(P_L - P_T) - \left[\sum_i \sigma_i P_i + \sum_j \delta_j \theta_j Z_j \right]. \qquad (2.16)$$

This first term in Eq. (2.16) is again the dilution equation for a conservative element, while the second term is the sum of phosphorus loss rates over all the phyto- and zooplankton populations. By inspection of Equation (2.2), it is seen that Eq. (2.16) is identical to the typical one-box total phosphorus loading model if we assume the identity of net sources and sinks term (S_P) with the terms inside brackets in Eq. (2.16):

$$S_P = \sum_i \sigma_i Q_i C_i + \sum_j \delta_j \theta_j Z_j. \qquad (2.17)$$

This relationship between the phosphorus cycling model and the total phosphorus mass balance is less trivial than it might seem: several models of nutrient cycling in the plankton community neglect the contribution of zooplankton to total nutrients, and thereby create serious problems in maintaining the necessary mass balances between nutrient intake, utilization, and regeneration in the grazer compartments. If zooplankton nutrient release is modeled merely by the allometric relationships between body size and release rate (Peters and Rigler 1973), and without any feedback from the actual nutrient intake via ingested food (e.g., Carpenter and Kitchell 1984), then zooplankton will actually be represented as nonstoichiometric entities with the ability to spontaneously create nutrient atoms.

2.6 Summary and Conclusions

The main features of common phosphorus mass-balance models based on the principles of Vollenweider (1968, 1976) have been briefly reviewed. It is shown that predictions from present models, based mainly on the hydraulic properties of lakes, contain ample amounts of residual variance both in the predicted lake total P content at a given P loading, and in the predicted algal biomass at a give total P level. It is proposed that both the retention of phosphorus and the yield of algal biomass can be viewed as results of biological processes deeply connected with the dynamics of the phosphorus cycle in the plankton community.

In an attempt to link a dynamic view of the pelagic phosphorus cycle with the mass-balance concepts of phosphorus loading models, a dynamic model is formulated which accounts for the partitioning of pelagic phosphorus between major compartments through the processes of uptake, grazing, and recycling. It is shown that this modeling framework is a valid augmentation to the single-compartment type of total phosphorus models. Although arguments of functions have been omitted in the model description above to increase readability, it must be kept in mind that, except for the parameters controlling the physical processes of dilution and sinking, all other parameters of the model are in reality nonlinear functions that are deeply interrelated with the state variables of the model. As it stands, the model developed in this chapter is still too general to be of much use for either analytical or numerical treatment; the next two chapters will therefore be devoted to further specification and parametrization of processes involved in controlling growth and nutrient utilization in phyto- and zooplankton populations.

2.7 Symbols, Definitions, and Units

Symbol	Definition	Dimension
C_i	Biomass of phytoplankton population i	$(\text{mg C}) \, l^{-1}$
δ_j	Mortality loss rate from zooplankton population j	day^{-1}
D	Dilution rate	day^{-1}
D_i	Total loss rate from phytoplankton population i	day^{-1}
F_{ij}	Clearance rate of zooplankton j on phytoplankton i	$l \, (\text{mg C})^{-1} \, \text{day}^{-1}$
g_i	Specific growth rate of zooplankton population j	day^{-1}
L_p	Total phosphorus loading	$(\mu g\, P) \, l^{-1} \, \text{day}^{-1}$
μ_i	Specific growth rate of phytoplankton population i	day^{-1}
q_i	Water flow into lake	$l \, \text{day}^{-1}$
q_0	Water flow out of lake	$l \, \text{day}^{-1}$
θ_i	Phosphorus content of zooplankton population j	$(\mu g\, P) \, (\text{mg C})^{-1}$
Q_i	Phosphorus content of phytoplankton population i	$(\mu g\, P) \, (\text{mg C})^{-1}$
P_I	Dissolved inorganic P concentration	$(\mu g\, P) \, l^{-1}$
P_T	Total phosphorus concentration in lake	$(\mu g\, P) \, l^{-1}$
P_L	Total phosphorus concentration in inflowing water	$(\mu g\, P) \, l^{-1}$
ρ_i	Phosphorus regeneration rate of zooplankton j	$(\mu g\, P) \, (\text{mg C})^{-1} \, \text{day}^{-1}$
R_L	Unavailable fraction of total P loading	–
R_P	Total phosphorus retention	–
σ_i	Sinking loss rate from phytoplankton population i	day^{-1}

Symbol	Definition	Dimension
σ_P	Net total P sedimentation loss rate	day^{-1}
S_P	Net internal total P supply or loss rate	$(\mu g\,P)\,l^{-1}\,day^{-1}$
V	Lake volume	l
v_i	Net P uptake rate in phytoplankton population i	$(\mu g\,P)\,(mg\,C)^{-1}\,day^{-1}$
u	Terminal sinking velocity	$m\,day^{-1}$
z_m	Depth of the mixed layer	m
Z_j	Biomass of zooplankton population j	$(mg\,C)\,l^{-1}$

3 Algae and Nutrients: Uptake and Utilization of Limiting Nutrients in Generalized Phytoplankton Species

> *In the steady state, uptake and growth are fully coupled, $v = \mu Q$, but that is not to imply that growth is controlled by uptake. On the contrary, the converse may apply...*
>
> M. R. Droop (1983)

In a nutrient-limited environment, a successful competitor must be able to maintain nonnegative growth at nutrient levels less than those required by other, competing species. The fate of a phytoplankton population is therefore to a large extent determined by the ability of its individual members to take up mineral nutrients from dilute solutions and the efficiency with which these nutrients are utilized to form new biomass. As phytoplankton constitute a major food source to grazing zooplankton, the stoichiometry of algal biomass will also affect the nutrient supply to higher trophic levels. Although similar principles seem to apply for a range of nutrients potentially limiting to phytoplankton growth (e.g., P, N, Fe, Mn; Morel 1987), the main focus in this chapter will be on the uptake and utilization of inorganic phosphorus, either as the single limiting nutrient or in a situation where both phosphorus and nitrogen may be limiting.

3.1 Growth and Nutrient Utilization

It is generally observed that the biomass yield in a nutrient-limited chemostat culture is negatively correlated to dilution rate (D; d^{-1}), and that this relationship is fairly linear. If we define the biomass carbon yield in a phosphorus-limited phytoplankton culture as Y [(mg C) (µg P)$^{-1}$], the steady-state yield can be written as

$$Y = Y'\left(1 - \frac{\mu}{\mu'}\right),$$
(3.1)

where we have incorporated the steady-state equivalence of dilution rate (D) and specific growth rate (μ; day^{-1}). The parameters of Eq. (3.1) are Y': the yield at zero growth rate [(mg C) (μg P)$^{-1}$], and μ': the growth rate corresponding to zero yield (day^{-1}). It must be noted that both the parameters of Eq. (3.1) are extrapolations to dilution rates where it is impossible to operate a chemostat culture.

If we reexpress Eq. (3.1) in terms of $Q = Y'$, the cellular phosphorus content (or P quota), and solve Eq. (3.1) with respect to μ, we obtain the familiar hyperbolic function introduced by Droop (1968):

$$\mu = \mu'\left(1 - \frac{Q'}{Q}\right).$$

(3.2)

Q' is often termed the subsistence P quota [(μ g P) (mg C)$^{-1}$], or the lower physiological limit for cellular phosphorus at which further cell division is impossible, while μ' (day^{-1}) is the asymptotic growth rate for Q approaching infinity. The two parameters in the expression (3.2) have been found sufficient to describe nutrient limited growth in a variety of species (e.g., Table 1 in Droop 1983). In a chemostat at equilibrium, causal relationships vanish, so that yield as a function of growth rate [Eq. (3.1)] and growth rate as a function of cell quota [Eq. (3.2)] are equivalent representations (Thingstad 1987). On the other hand, the Droop equation has also been shown to be applicable to describe nutrient-limited growth in several natural communities (Goldman et al. 1979; Olsen et al. 1983b; Sommer 1988a, 1989a).

As organisms have a finite storage capacity for nutrients, it is obvious that, for a phytoplankton population growing at given levels of light and temperature, the specific growth rate must be constrained below a maximum rate μ'' (day^{-1}) with the property that $\mu'' < \mu'$. Corresponding to the maximum growth rate μ'', we can define a maximum cell quota Q'' [(μg P) (mg C)$^{-1}$] so that substituting $\mu = \mu''$ and $Q = Q''$ into Eq. (3.2) and rearranging gives the following relationship between the parameters μ', μ'', Q', and: Q''

$$\frac{\mu''}{\mu} = 1 - \frac{Q'}{Q''}.$$

(3.3)

This means that if the specific growth rate is constrained to the range $0 \le \mu \le \mu''$, then the P quota must remain within $Q' \le Q \le Q''$ if Eq. (3.2) is to be valid. Any one of the four parameters μ', μ'', Q', and Q'' can be found by solving Eq. (3.3) if the three remaining are known. If we define the cellular storage capacity as Q''/Q', it is evident that a large storage capacity means near equivalence of μ' and μ'', while a small storage capacity implies that μ' is significantly larger than μ''. The storage capacity will to a large extent be determined by the bulk cell content of a given nutrient, with large Q''/Q' in microconstituents like vitamin B12 (Droop 1968) and small Q''/Q' in macroconstituents like nitrogen (Goldman and McCarthy 1978).

3.2 Nutrient Uptake

Plankton algae generally obtain their constituent mineral nutrients from the surrounding water. Only in rare cases are all essential nutrients present in the water in concentrations comparable to those that must be maintained within the cell. This means that nutrients must be actively transported into the cell against a concentration gradient, and that this concentration gradient gives rise to a passive diffusive flux out of the cell. It has thus been shown that nutrient uptake in microorganisms can be described as the net result of fast, bidirectional exchange processes across the cell membrane (Lean 1976; Lean and Nalewajko 1976; Olsen 1989). While some authors argue that only the net uptake rate is of importance to phytoplankton growth (e.g., Lehman 1984), other studies have shown that the fast exchange processes can be very important to the outcome of competition under certain modes of nutrient supply (Olsen et al. 1989).

In short-term uptake measurements, it is generally found that the gross nutrient uptake rate is a nondecreasing function of dissolved nutrient concentration, which levels off to a saturated uptake rate at high nutrient levels. On a longer time scale there is found to be a feedback interaction between uptake and growth such that general response to nutrient limitation seems to be an increase in the saturated uptake rate (see Turpin 1988 and references therein). While the Michaelis-Menten model of enzyme kinetics is often used to describe cellular nutrient uptake, it should be noted that several workers have observed deviations from this model, indicating the existence of two or more uptake systems with different substrate affinities (Brown et al. 1978; Brown and Button 1979; Olsen 1989). At low concentrations of dissolved nutrient, the relationship between uptake rate and external concentration will nevertheless be dominated by the uptake system with the highest substrate affinity and closely approximated by a linear function. Olsen (1989) presented evidence that the information contained in such a first-order approximation was sufficient to predict the outcome of chemostat competition experiments. Apart from the mathematical convenience, it can be argued in defense of using first-order kinetics instead of the more common saturation kinetics, that nutrient concentrations will almost by definition be low as long as there is competition for nutrients in the plankton community, and that when a nutrient is nonlimiting to all members of the community it does not really matter how the uptake kinetics are described.

If we consider the case of inorganic phosphorus uptake, the first-order approximation implies that the net, carbon-specific phosphorus uptake rate v [(μg P) (mg C)$^{-1}$ day^{-1}], which is the difference between an influx rate (v_i) and efflux rate (v_e), can be written as

$$v = v_i - v_e = \alpha(S - S'), \qquad (3.4)$$

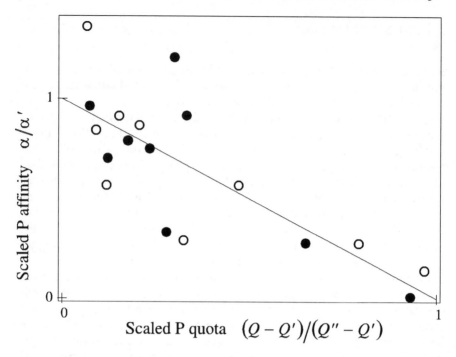

Fig. 3.1. Inorganic P affinity as function of cellular P content with both axes scaled to dimensionless quantities suggested by Eq. (3.5). *Open symbols: Staurastrum luetkemuellerii,* solid symbols: *Microcystis aeruginosa; solid line* predicted relationship from Eq. (3.5). (After Olsen 1989)

where S is the ambient inorganic P concentration and S' is the critical concentration where net uptake rate becomes zero [both in units of $(\mu g\ P)\ l^{-1}$]. α [l $(mg\ C)^{-1}\ day^{-1}$] is the carbon biomass specific substrate affinity of the inorganic P uptake system, which will be equal to the slope of the relationship between uptake rate and concentration extrapolated to zero concentration (Button 1978). The affinity is given in the same unit as the clearance rate of a filter-feeding grazer (see Sect. 2.5), pointing to a conceptual relatedness between the two parameters; both can be interpreted as the virtual volume of water which will have all its content of some substance (either nutrient ions or food particles) removed by the activity of a unit biomass in a unit of time.

Several studies (e.g., Perry 1976; Gotham and Rhee 1981; Riegman and Mur 1984; Olsen 1989) have shown that the typical response to phosphorus limitation is an increase in the saturated P uptake rate. Morel (1987) has shown that experimental data from several algal species and different limiting nutrients are consistent with a linear relationship between saturated uptake rate and cell quota. Restated in terms of the first-order approximation (3.4), this relationship can be written as

$$\alpha = \alpha' \frac{Q' - Q}{Q'' - Q'}, \tag{3.5}$$

where α' is the maximum inorganic P affinity corresponding to zero growth rate. Within the attainable precision of current affinity measurement methodology, a more complex model than the linear function (3.5) seems unwarranted (Fig. 3.1).

For simulation purposes, a few extra precautions should be taken to avoid some rather annoying properties hidden in Eqs. (3.4) and (3.5). From Eq. (3.5) it is seen that the affinity α becomes negative when $Q > Q''$. As long as $S < S'$, this is a desirable property, which ensures that Q cannot exceed Q''. When $Q > Q''$ and $S < S'$, we have the undesirable property that v becomes positive and even increases as S decreases. A simple remedy for this is to modify the model such that v is set to zero whenever $Q > Q''$ and $S < S'$. In a similar fashion, one should ensure that $Q \geq Q'$ (and accordingly $\mu \geq 0$) also when $S < S'$ by setting $v = 0$ whenever $Q < Q'$ and $S < S'$. The last modification can be justified by assuming that the minimal amount of cellular P (Q') is a constitutional quantity that cannot be lost through diffusive efflux processes (Shuter 1978).

3.3 Balancing Nutrient Uptake and Growth

Under steady-state conditions where the algal growth rate is exactly balanced by a constant loss rate, the conservation of mass requires that net nutrient uptake must be equal to algal production expressed in nutrient units. This requirement is usually written as

$$v = \mu Q, \tag{3.6}$$

or for the specific case of P-limited growth, that the net, specific inorganic P uptake rate equals specific growth rate times P quota (Burmaster 1979; Droop 1983; Morel 1987). By solving Eq. (3.2) with respect to Q and substituting into Eq. (3.5), the inorganic P affinity [Eq. (3.6)] can be reexpressed as

$$\alpha = \alpha' \frac{\mu'}{\mu''} \frac{\mu'' - \mu}{\mu' - \mu}. \tag{3.7}$$

Likewise, the Droop equation [Eq. (3.2)] can be rearranged such that the right-hand side of Eq. (3.6) is expressed as a function of μ alone

$$\mu Q = \frac{\mu' \mu}{\mu' - \mu} Q'. \tag{3.8}$$

Substituting Eqs. (3.4), (3.7), and (3.8) into Eq. (3.6) leads to the following steady-state relationship between inorganic P concentration and specific growth rate

$$\alpha' \frac{\mu'}{\mu''} \frac{\mu'' - \mu}{\mu' - \mu} (S - S') = \frac{\mu' \mu}{\mu' - \mu} Q'. \tag{3.9}$$

After eliminating common factors, this can be rearranged to a form which, for the case of negligible nutrient efflux $(S' = 0)$, is identical to the Monod model of nutrient-limited growth

$$\mu = \mu'' \frac{(S - S')}{K' + (S - S')}. \tag{3.10}$$

The half-saturation parameter for P-limited growth, K' $[(\mu g \ P) \ l^{-1}]$, is given by

$$K' = \frac{\mu'' Q'}{\alpha'}. \tag{3.11}$$

This equivalence of growth models based on the Droop equation [Eq. (3.2)] and the Monod model under steady-state conditions has been proven several times before for different assumptions on nutrient uptake kinetics (Burmaster 1979; Droop 1983; Morel 1987). For example, both Burmaster (1979) and Morel (1987) formulated models based on the common assumption that the nutrient uptake rate can be described by the Michaelis-Menten model. Changing their notation to conform with the rest this chapter, the Michaelis-Menten model can be written as

$$v = v_m \frac{S}{K_m + S}, \tag{3.12}$$

where v_m is the saturated uptake rate (approached asymptotically as $S \to \infty$) and K_m is the half-saturation parameter ($S = K_m$ implies that $v = v_m/2$). Burmaster (1979) assumed no feedback between nutrient status and nutrient uptake, so that both the parameters v_m and K_m are considered constant. Under the steady-state condition [Eq. 3.6] one can proceed as above by eliminating Q in Eqs. (3.2), (3.6), (3.12) and solving for μ. This gives the following Monod expression

$$\mu = \mu'' \frac{S}{K_m \dfrac{\mu'' Q}{v_m} + S}, \tag{3.13}$$

which is identical to Eq. (3.10) for the case of $S' = 0$, if we substitute the definition of affinity as the initial slope of Michaelis-Menten curve (i.e., $\alpha = v_m/K_m$).

In an extension to the model of Burmaster (1979), Morel (1987) introduced feedback between nutrient status and uptake rate. He assumed that the regulation of nutrient uptake is effected through the saturated uptake rate v_m, while the half-saturation parameter K_m can be considered constant. In his model, v_m is a linear decreasing function of cell quota with a maximal saturated uptake rate v'_m at maximal nutrient starvation ($Q = Q'$) and a minimal saturated uptake rate v''_m corresponding to ($Q = Q''$) That is,

$$v_m = v'_m - (v'_m - v''_m)\frac{Q - Q'}{Q'' - Q'}. \qquad (3.14)$$

Morel (1987) showed that under the steady-state condition [Eq. (3.6)] this model also is equivalent to a Monod-type relationship between growth rate and external concentration, which can be written as

$$\mu = \frac{v''_m}{Q''} \frac{S}{K_m \dfrac{v''_m Q'}{v'_m Q''} + S}. \qquad (3.15)$$

If we take into consideration the definition of affinity ($\alpha' = v'_m / K_m$) and that the steady-state condition [Eq. (3.6)] implies that $v''_m = \mu'' Q''$ when $\mu = \mu''$, it is clear that Eq. (3.15) is also the equivalent of Eq. (3.10) for the case where nutrient efflux is negligible ($S' = 0$).

3.4 Balancing Photosynthesis and Growth

In a particularly elegant, but for some reason rarely cited, model of balanced phytoplankton growth, Shuter (1979) assumed that the material in algal cells can be partitioned into four major compartments: structural, photosynthetic, biosynthetic, and storage. The structural compartment was assumed to be a fixed fraction of cell mass, while the other three were allowed to vary in response to the growth conditions. Shuter (1979) showed that if the growth rate is constrained by a given level of light or nutrient limitation, there will be a unique, optimal allocation of photosynthate into the three variable compartments, that will ensure balanced growth. In this context, balanced growth means that cellular proportions between the four compartments will remain fixed from generation to generation.

If growth is nutrient-limited, the model of Shuter (1979) predicts that the photosynthetic compartment should increase in proportion to the growth rate. This means that the cell should not allocate more material to the photosynthetic apparatus than is needed to produce the amount of photosynthate required to support growth. In the limiting case where nutrient limitation is so severe than no net growth is possible, the size of the photosynthetic compartment should only be sufficient to support maintenance.

On the other hand, if growth is light-limited, Shuter (1979) predicts that the photosynthetic compartment should decrease with increasing growth rate. Both light- and nutrient-limited growth should converge to the same allocation rule as growth becomes unlimited. The model predictions by Shuter (1979) been verified experimentally by, among others, Laws et al. (1983, 1985)

If the photosynthetic compartment has a fixed chla:C ratio, while the other compartments contain no photosynthetic pigments, then the chla:C ratio of the whole cell will be proportional to the relative size of this compartment. If we denote the cellular ratio by φ [(μg chla) (mg C)$^{-1}$], the relationship between chlorophyll a content and nutrient-limited growth rate (μ) can be written as

$$\varphi = \varphi' \frac{\mu + \mu_0}{\mu' + \mu_0}, \qquad\qquad (3.16)$$

where μ_0 (day^{-1}) is the maintenance growth rate (sensu Geider et al. 1985) and φ' is the chla:C ratio for unlimited growth ($\varphi \to \varphi'$ when($\mu \to \mu'$). The chlorophyll a concentration will be given by φC, which implies that the same chla level could result equally well from a slowly growing population with high biomass, as from a fast-growing population with low biomass.

3.5 Phytoplankton growth model parametrization

Natural phytoplankton communities contain species from at least nine taxonomic classes, of which several hundred species can be commonly encountered (depending on the taxonomic ability of the investigator). A minority of these species have so far been studied in pure culture (several common species have so far proven unculturable), and in only a few of these studies have all parameters of the model presented in Sections 3.1 and 3.2 been measured simultaneously. Thus, any model parametrization specific to a representative collection of common species seems so far impossible; one can only hope to obtain a rough indication of parameter location and variability within the somewhat arbitrary set of species from which estimates are available. Given these constraints, the present section contains a literature survey of data on the five parameters that are considered fundamental to the processes of phytoplankton growth and nutrient utilization under phosphorus limitation: maximum growth rate (μ''), minimum and maximum cell quota (Q' and Q''), maximum nutrient uptake affinity (α'), and threshold nutrient concentration for positive net uptake (S'). Since algal biomass so often is reported as chlorophyll a concentration in field investigations, we will also consider the parametrization of the carbon:chlorophyll a relationship [Eq. (3.16)] in terms of the chla:C ratio for unlimited growth (φ').

Maximum Growth Rate (μ''). Several workers (Sommer 1983; Reynolds 1989) have reported that plankton algae conform to the general allometric relationships between size and growth rate (e.g., Blueweiss et al. 1978; Peters 1983), in that small-celled species are found to be able to grow faster than large-celled ones. On the other hand, both Banse (1982) and Furnas (1990) point out that there is also important variation related to the morphological features of the major taxonomic groups; marine diatoms have significantly higher growth rates than would be expected from their size, while the reverse seems to be true for marine dinoflagellates. Within both of these groups, there are only weak relationships between size and growth rate (Banse 1982).

A reasonably large amount of data exist on maximal growth rates of plankton algae (Table A10.1). The cumulative frequency distribution of the data set (Fig. 3.2) is quite symmetrical, with some resemblance to a truncated normal distribution. The median μ'' is 1.2 day^{-1} with about twofold variation within the central 50% of the distribution (interquartile range of 0.8-1.8 day^{-1}).

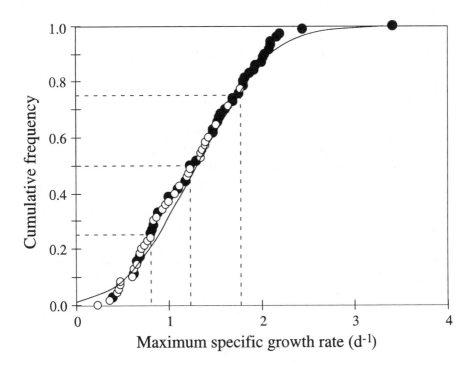

Fig. 3.2. Cumulative frequency distribution of maximum specific growth rates (day^{-1}) at 20 °C in different species of plankton algae (data from Table A10.1; $n = 71$). *Solid line* Fitted normal distribution; *broken lines* median and upper and lower quartiles; *open symbols* grazing-resistant species; *solid symbols* nongrazing-resistant species

As cell size is a major determinant of edibility to zooplankton (e.g., Sterner 1989, and references therein), one would expect small species to need a higher maximal growth rate in order to compensate for higher grazing losses. The classification adopted here follows the current consensus with regards to cladoceran feeding preferences as outlined by Sterner (1989): grazing-resistant species are either large (equivalent spherical cellular or colonial diameter > 20 μm), or elongated (largest axial dimension > 100 μm), or have specialized morphological adaptions like long spines or a resistant gelatinous sheath. Within this classification, the grazing-resistant group has significantly lower mean growth rate than the nongrazing-resistant group (t = 4.87, n = 71, p < 0.0001). The median growth rates of the two groups (resistant: 0.9 day^{-1}, nonresistant: 1.7 day^{-1}) are such that the lower and upper quartiles in Fig. 3.2 might also be taken to represent algae with high and low resistance to zooplankton grazing.

The extremes within the two distributions are represented by diatoms which have generally higher growth rates than most other grazing-resistant algae, and cryptomonads which have some of the lowest growth rates among the grazing-susceptible species. The relatively high growth rates of large freshwater diatoms fit well with the pattern observed by Furnas (1990) in marine plankton algae. Comparatively low growth rates in cryptomonads are puzzling when taking into account the commonly observed dominance of this group under conditions of very heavy grazing. Heaney and Sommer (1984) indicate that cryptomonads might not reach their open-water growth rates when enclosed in bottles, and that cryptomonad growth rates therefore might be biased by some containment effect.

Minimum P Cell Quota (Q'). Shuter (1978) argued from a review of published phosphorus subsistence quotas that there is an allometric relationship between cell size and the minimal amount of cellular P, so that small cells would be expected to contain more P per unit carbon or volume than large cells. In the light of the recent findings of very high P contents in heterotrophic bacteria (e.g., Vadstein et al. 1988), one might suspect the conclusions of Shuter (1978) to be influenced by the inclusion of several species of bacteria in his data set. If bacteria are excluded from the data set of Shuter (1978), regression analysis on log-transformed cell volumes and cellular P contents gives a slope of 0.92 ± 0.10, when corrected for bias by the geometric means method recommended by Ricker (1973). As the slope is not significantly different from 1, this reanalysis of the data set compiled by Shuter (1978) does not support the idea of size-dependent phosphorus subsistence quotas in autotrophic microorganisms.

Table A10.2 lists the data used by Shuter (1978) recalculated to carbon-specific subsistence quotas together with several more recent accounts on minimal P requirements in species of plankton algae. Analysis of variance on the data in Table A10.2 indicated no significant differences either between prokaryote and eukaryote autotrophs (ANOVA: $F_{1,40}$ = 2.31, p = 0.137),

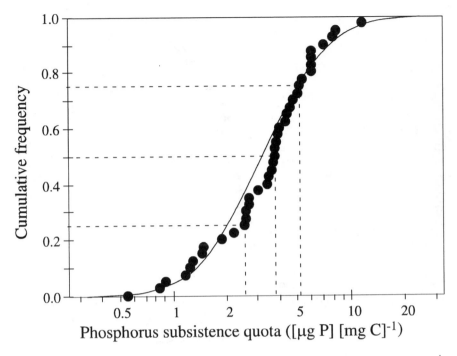

Fig. 3.3. Cumulative frequency distribution of phosphorus subsistence quotas [(μg P) (mg C)$^{-1}$] in different species of plankton algae (data from Table A10.2; $n = 40$).). *Solid line* Fitted lognormal distribution; *broken lines* median and upper and lower quartiles

nor between marine and freshwater species (ANOVA: $F_{1,40} = 0.014$, $p = 0.908$). The resulting frequency distribution when all data in Table A10.2 are pooled (Fig. 3.3), indicate that phosphorus subsistence quotas are approximately lognormally distributed among species with median 3.8 (μg P) (mg C)$^{-1}$ and a twofold interquartile range of 2.5-5.2 (μg P) (mg C)$^{-1}$.

Maximum Cell Quota (Q ″). The minimum and maximum P quotas are not independent parameters in the sense that we must always have $Q' \le Q''$. Given the variability in Q' as shown in Fig. 3.3, the distributions of Q' and Q'' when treated as independent parameters will most certainly be overlapping. In order to ensure that $Q' \le Q''$, it should be safer to estimate Q'' indirectly from the distribution of the phosphorus storage capacity, defined as the ratio of maximal to minimal P requirements, or Q''/Q'. This approach also allows the utilization of data sets where P quotas are given on a per-cell basis (e.g., Gotham and Rhee 1981) without any conversion factors to cell volume or carbon. On the other hand, several studies describing P-limited growth through the Droop model have not included the information necessary to calculate the ratio Q''/Q', so that the data set on phosphorus storage capacities (Table A10.3) is less extensive ($n = 20$)

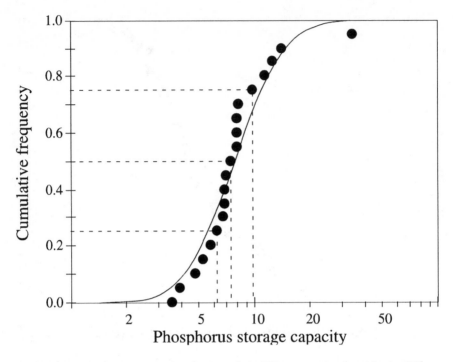

Fig. 3.4. Cumulative frequency distribution of phosphorus storage capacities in different species of plankton algae (data from Table A10.3; $n = 20$). *Solid line* Fitted lognormal distribution; *broken lines* median and upper and lower quartiles

than the data sets on subsistence quotas and growth rates. The frequency distribution of the ratio Q''/Q' (Fig. 3.4) is skewed to the right, somewhat resembling a lognormal distribution, with median 7.5 and interquartile range from 6.3 to 9.7. For the median subsistence quota $[Q' = 3.8$ (µg P) (mg C)$^{-1}]$ a typical value $Q'' = 28.5$ (µg P) (mg C)$^{-1}$ is suggested, which is slightly above the Redfield ratio [atomic C:P = 106:1, or 24.4 (µg P) (mg C)$^{-1}$). Substituting the median Q''/Q' into Eq. (3.3) further indicates that for P-limited growth we can assume $\mu' \approx 1.15\ \mu''$.

Maximum Phosphorus Uptake Affinity (α'). As many workers have reported their measurements of P uptake kinetics on a per-cell basis, many of the published data are not readily available in biomass-specific units, so that interspecific comparisons can be done. If we define the relative uptake affinity as $(\alpha'/Q'$ [l (µg P)$^{-1}$ day^{-1}], we have a quantity which can be calculated from data reported in both cell-specific and biomass-specific units. By inspection of Eqs. (3.11), (3.13), and (3.15), it is seen that the relative uptake affinity can also be estimated from the parameters of the Monod model as

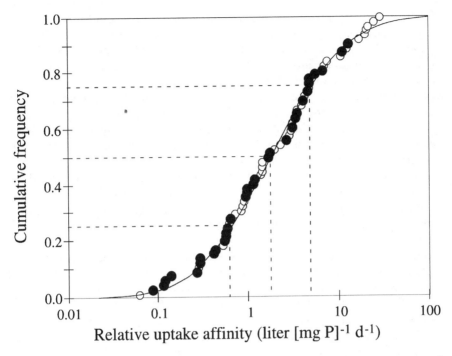

Fig. 3.5. Cumulative frequency distribution of relative phosphorus uptake affinities (l (μg P)$^{-1}$ day^{-1}) in different species of plankton algae (data from Table A10.4; $n = 64$). *Solid line* Fitted lognormal distribution; *broken lines* median and upper and lower quartiles; *open symbols* grazing-resistant species; *solid symbols* nongrazing-resistant species

$$\frac{\alpha'}{Q'} = \frac{\mu''}{K'}. \qquad (3.17)$$

The comparatively large body of P-limited growth data fitted to the Monod model can therefore also be used to estimate the relative phosphate uptake affinity α'/Q'.

Figure 3.5 shows that the data compiled in Table A10.4 span more than two orders of magnitude with median 1.7 l (μg P)$^{-1}$ day^{-1} and interquartile range from 0.6 to 4.8 l (μg P)$^{-1}$ day^{-1}. As was the case with the parameters Q' and Q''/Q'., the distribution is skewed to the right with a reasonably good fit to a lognormal distribution. Several authors (e.g., Malone 1980) have proposed that small algal cells should have a more efficient nutrient uptake than large cells, thus we should expect the P uptake affinity to be lower in large, grazing-resistant species than in the nonresistant species.

Figure 3.5 does not indicate a similar distinction between species with different grazing resistance as in the maximum growth rate data (Fig. 3.2). In fact, the trend in Fig. 3.5 is slightly toward higher affinity in the grazing-resistant species, although not significant ($t = 1.19$, $n = 64$, $p = 0.24$). This result is perhaps not so surprising when taking into account the variety of experimental conditions and analytical methods used by the different authors. The half-saturation parameter of the Monod model (K') is dependent on accurate inorganic P measurements at concentrations close to or below the detection limit of standard chemical methods. This means that the presence of a threshold concentration (S') as in Eq. (3.10) would easily be undetected, leading to an overestimation of K', thus tending to underestimate α'/Q'. On the other hand, direct measurement of uptake parameters can also be seriously biased if, for example, the possible existence of multiple uptake systems is neglected. Olsen (1989) found that in the green algae *Staurastrum luetkemuellerii* P affinity measurements at low concentrations [< 4 (μg P) l^{-1}] were at least twice as high as measurements made at concentrations more commonly used in P uptake experiments, due to the presence of a high-affinity uptake system in this species.

   Even if we attribute a major part of the variance in Fig. 3.5 to methodological aspects, it still seems reasonable to conclude that the biological variability in the relative uptake affinity is large compared to the variability in other model parameters like the P subsistence quota. For the median P subsistence quota [$Q' = 3.8$ (μg P) (mg C)$^{-1}$] a typical value $\alpha' = 6.5$ l (mg C)$^{-1}$ day^{-1} is suggested, while values corresponding to lower and upper quartiles of the distribution [2.3 and 18.2 l (mg C)$^{-1}$ day^{-1}] might be considered representative for species with particularly low and high inorganic P affinity.

Threshold Concentration for Positive Net Uptake (S'). Compared to the relative abundance of net or gross nutrient influx measurements, very few studies have attempted to measure nutrient effluxes from phytoplankton. Due to the close coupling of the influx and efflux processes, such measurements can be made only indirectly through tracer kinetics (Lean and Nalewajko 1976) or mass-balance calculations (Olsen 1989). Only the data set of Olsen (1989) contains simultaneous measurements of biomass-specific P efflux rates [v_e ; (μg P) (mg C)$^{-1}$ day^{-1}] and inorganic P uptake affinities, such that the threshold concentration for positive net uptake (S') can be calculated. From the net uptake equation [Eq. (3.4)] we observe that the the efflux rate can be written as $v_e = \alpha S'$, and that this relationship can be used to calculate S' when α and v_e are known. From the data presented in Olsen (1989), we can calculate that $S' = 0.07$- 0.60 (μg P) l^{-1} for *Microcystis aeruginosa* and 0.07-0.35 (μg P) l^{-1} in *Staurastrum luetkemuellerii*, when corrected for the presence of a high-affinity uptake system. Although admittedly based on limited data, one might conclude that the threshold concentration for positive net uptake (S') is in the order of a few tenths of a μg P per liter, and thus located safely below the detection limit of standard chemical methods. It must nevertheless be emphasized that the smallness in magnitude of this parameter does not necessarily reflect its significance in determining the outcome of algal nutrient competition (Olsen 1989; Olsen et al. 1989).

The Chlorophyll : Carbon Ratio. The maintenance costs appear to be quite low in microalgae, with many estimates being not significantly different from zero (Geider 1987). A fairly reasonable value for the maintenance rate in Eq. (3.16) seems to be $\mu_0 = 0.05$ day^{-1}, corresponding to the maintenance rate found by Laws et al. (1983), for both light- and nutrient-limited cultures of *Thalassiosira weisflogii*. According to the theory of Shuter (1979), the maximum chla:C ratio for nutrient-limited growth (φ'), could equally well be estimated as the minimum chla:C ratio for light-limited growth. Table A10.5 shows a collection of φ' estimates from different phytoplankton species, obtained under either light- or nutrient-limited growth. The φ' values in Table A10.5 vary by more than an order of magnitude, with the lowest value found in the slowly growing marine dinoflagellate *Prorocentrum micans*. The median of the data in Table A10.5 is 18 (µg chla) (mg C)$^{-1}$ with an interquartile range from 12 to 20 (µg chla) (mg C)$^{-1}$.

Properties of Generalized Phytoplankton Species. It is recognized that the diversity of experimental conditions underlying the individual entries in Tables A10.1 to A10.4 is a potential source of bias in the parameter distributions, and that such bias is most likely to affect the extremes of the distribution. A more robust measure of parameter variability would be to consider the upper and lower quartiles of the distribution, which can also be interpreted as the median of the values below the median and the median of the values above the median. If we assign the categories low, typical, and high to the first, second, and third quartiles of each distribution, as in Table 3.1, we have the foundation for constructing a range of contrasting "species" with different competitive advantages under P-limitation, while at the same time remaining within realistic parameter values.

Table 3.1. Summary of phytoplankton uptake and growth parameters. Parameter values termed low, typical, and high correspond to lower quartile, median and upper quartile of the frequency distributions shown in Figs. 3.2-3.5.

	Parameter	Low	Typical	High	Unit
μ''	Maximal growth rate	0.8	1.2	1.8	day^{-1}
Q'	Subsistence P quota	2.5	3.8	5.2	(µg P) (mg C)$^{-1}$
Q'/Q''	P storage capacity	6.3	7.5	9.7	-
μ'/μ''	Rel. asymptotic growth rate	1.11	1.15	1.19	-
α'/Q'	P-specific P uptake affinity	0.6	1.7	4.8	l (µg P)$^{-1}$ day^{-1}
α'	C-specific P uptake affinity	2.3	6.5	18.2	l (mg C)$^{-1}$ day^{-1}

3.6 Resource Competition Under Phosphorus Limitation

The r–K strategy tradeoff, a concept first introduced by MacArthur and Wilson (1967) in terms of the parameters of the logistic growth equation, has been proposed as a useful classification for phytoplankton species with different competitive traits (Kilham and Kilham 1980; Sommer 1981; Reynolds 1984; Kilham and Hecky 1988). Typical r-strategists are opportunist species with high maximal growth rates, and thus high capacities for exploiting and monopolizing sudden pulses of resource abundance. By contrast, typical K-strategists are efficient exploiters of situations where resources are uniformly scarce and competition is high. For such a classification to be meaningful there must be a tradeoff between the two traits so that no species can be competitively superior at all levels of resource availability.

The quartile statistics of phytoplankton growth and uptake parameters in Table 3.1 is a reasonable starting point for constructing realistic model species exhibiting a tradeoff between r- and K-strategies. Such model species should be regarded as extremes of what could be encountered in nature, rather than as mimicking any particular phytoplankton species. If we assume an inverse relationship between the maximal growth rate (μ'') and the maximal phosphorus uptake affinity [α'; l (mg C)$^{-1}$ day^{-1}], we can construct an r–K gradient composed of the three species listed in Table 3.2. The phosphorus subsistence quota [Q'; (μg P) (mg C)$^{-1}$], and the ratios Q''/Q' and μ'/μ'' are identical in all three species, and equal to the median values of Table 3.1. At this stage, we will for simplicity assume that there is no efflux of phosphorus from any of the species (that is, the threshold level for positive net P uptake, S', is zero). From these parameter values the corresponding Monod half-saturation parameter (K'), given by Eq. (3.11), will range from 0.17 to 3.0 (μg P) l^{-1} in the three species.

Table 3.2. Parameters describing three idealized phytoplankton species with a tradeoff between r- and K-strategies, constructed from the quartile statistics in Table 3.1

	Parameter	1	2	3	Unit
μ''	Maximal growth rate	0.80	1.20	1.80	day^{-1}
μ'	Asymptotic growth rate	0.92	1.38	2.08	day^{-1}
Q'	Subsistence P quota	3.8	3.8	3.8	(μg P) (mg C)$^{-1}$
Q''	Maximal P quota	28.5	28.5	28.5	(μg P) (mg C)$^{-1}$
α'	Maximal P uptake affinity	18.2	6.5	2.3	l (mg C)$^{-1}$ day^{-1}
K'	Monod parameter	0.17	0.70	3.0	(μg P) l^{-1}

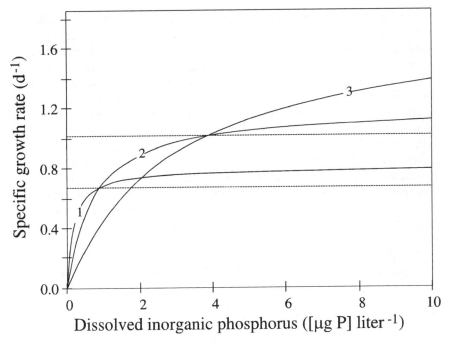

Fig. 3.6. Monod curves (numbered solid lines) relating steady state growth rate to equilibrium DIP concentration for the three model species in Table 3.2. *Dashed horizontal lines* bound the range of steady-state growth rates where a given species is competitively superior

From the parameters in Table 3.2 we can construct Monod curves [Eq. (3.10)], describing the relationship between specific growth rate and the equilibrium concentration of dissolved inorganic phosphorus when phosphorus uptake and growth are in balance. The relative positions of the Monod curves determine the competitive ability of phytoplankton species that are subject to a constant loss rate [which in the absence of grazing and sedimentation will be equal to the dilution rate (D; day^{-1}) of the system]. The equilibrium inorganic P level for a single phytoplankton species growing at equilibrium with dilution losses is given by substituting $\mu = D$ in the Monod equation [Eq. (3.10)] and solving for S. The species with the lowest equilibrium resource level will be able to invade any community dominated by inferior competitors to the extent of competitive exclusion, while the new equilibrium will be uninvadable to the inferior competitors (Hsu et al. 1977; Tilman 1982). For the set of the model species (Fig. 3.6), this means that equilibria of species 1 will be uninvadable at low loss rates, while in a gradient of increasing loss rates, species 1 will be successively replaced by species 2, which again will be replaced by species 3.

Phosphorus Competition Between Algae and Bacteria. Although planktonic bacteria have traditionally been considered as sources of inorganic nutrients through their activity as mineralizers of dissolved organic material, recent research has indicated that growth of heterotrophic bacteria may often appear to be limited by inorganic nutrients rather than organic carbon (Azam et al. 1983; Thingstad 1987). It will therefore be of some importance to make an assessment of the relative competitive abilities of these two functional compartments of the pelagic food web. Vadstein et al. (1988) found that bacteria have variable internal stores of phosphorus, and that the relationship between growth rate and P:C ratio could be described by the Droop model, much in the same way as with algae. Data on phosphorus-limited growth in bacteria are generally less abundant than for algae, but published values give a fair indication that the minimal P requirements of bacteria are an order of magnitude higher than in algae. Bacterial P subsistence quotas [Q_b'; $(\mu g\ P)\ (mg\ C)^{-1}$] compiled by Shuter (1978) and Vadstein et al. (1988) range from 17 to 72, with a median around 40.

As might be expected from the high minimum requirements for phosphorus found in bacteria, their storage capacity seem to be limited compared to what is observed in algae. For example, the data of Chen (1974) indicate $Q_b''/Q_b' = 2.2$ in *Corynebacterium bovis*, while observations from a natural community in the lake Nesjøvatn (Vadstein et al. 1988) indicate $Q_b''/Q_b' = 3$ (when omitting one high value that also was excluded in their fit to the Droop hyperbola). Assuming a phosphorus storage capacity in bacteria around 3 translates the asymptotic growth rate $\mu_b' = 0.13\ h^{-1} = 3.1\ day^{-1}$ estimated in both Vadstein et al. (1988) and Vadstein and Olsen (1989), to a maximal growth rate of $\mu_b'' = 2.1\ day^{-1}$.

The relatively few data available (summarized in Table 5 of Vadstein and Olsen 1989) indicate that maximal P-specific uptake affinities in planktonic bacteria are at least an order of magnitude larger than in phytoplankton, suggesting a typical value around $\alpha_b'/Q_b' = 40\ l\ (\mu g\ P)^{-1}\ day^{-1}$ for bacterio-plankton. Combining this with a subsistence quota of 40 $(\mu g\ P)\ (mg\ C)^{-1}$ gives a maximal P uptake affinity of 1600 l $(mg\ C)^{-1}\ day^{-1}$, or more than two orders of magnitude higher than the parameter value considered typical for algae in Table 3.1.

Table 3.3. Phosphorus uptake and utilization parameters in the model representations of algae and bacteria

	Parameter	Algae	Bacteria	Unit
μ''	Maximal growth rate	1.2	2.1	day^{-1}
μ'	Asymptotic growth rate	1.4	3.1	day^{-1}
Q'	Subsistence P quota	3.8	40	$(\mu g\ P)\ (mg\ C)^{-1}$
Q''	Maximal P quota	28.5	120	$(\mu g\ P)\ (mg\ C)^{-1}$
α'	Maximal P uptake affinity	6.5	1600	$l\ (mg\ C)^{-1}\ day^{-1}$
K'	Monod parameter	0.70	0.05	$(\mu g\ P)\ l^{-1}$

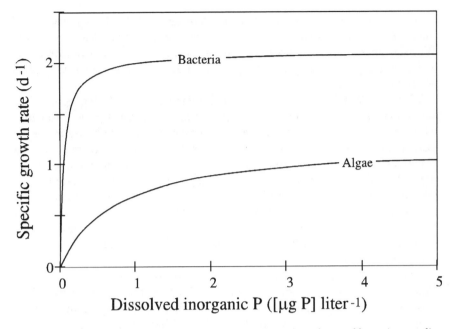

Fig. 3.7. Monod curves describing phosphorus limited growth in algae and bacteria according to the parameter sets given in Table 3.3

Table 3.3 summarizes typical phosphorus uptake and utilization parameters for bacteria compared to corresponding parameters for algae, taken from Table 3.1. The resulting Monod half-saturation parameters for typical algae and bacteria [given by Eq. (3.11)] differ by more than an order of magnitude, making phosphorus-limited bacteria competitively superior to algae (Fig. 3.7). This competitive advantage of bacteria would be maintained even if we had chosen the most favorable algal parameter combination from Table 3.1 [that is, using the high values of μ'' and α', and the low value of Q' in Table 3.1 gives $K' = 0.25$ (µg P) l^{-1}, which is still higher by a factor of 5 than the bacterial half-saturation parameter in Table 3.3].

3.7 More Than One Limiting Nutrient – P and N Limitation

Although some authors have advocated the idea that multiple limiting factors affect phytoplankton growth in a multiplicative fashion (Rodhe 1978; Bader 1982), it seems now generally accepted that a threshold model is more appropriate for describing situations where more than one essential element can become limiting to phytoplankton growth (Tilman 1982;

Droop 1983). The threshold model is equivalent to the Liebig law of the minimum in that growth is limited by the single element least in supply. Whether an essential resource becomes growth-limiting to a given organism will then depend on the relative supply rate of this resource compared to other essential resources.

Tilman (1980) introduced a graphical method for analyzing equilibrium situations with two limiting resources based on the zero net growth isoclines, which are the set of equilibrium resource levels where growth and losses are exactly balanced, and thus net growth is zero. For the case of two essential resources, the zero net growth isocline will consist of two straight line segments, parallel to the resource axes, and intersecting at right angles at a point where both resources are equally limiting (Fig. 3.8). For a pair of species to be able to coexist at fixed densities on two essential resources, the isoclines must intersect; that is, none of the species can be competitively superior for both resources. At the intersection point, each species will be limited by the resource on which it is competitively inferior.

Tilman (1980) has shown that the intersection of the zero net growth iso-clines is a necessary, but not sufficient, condition for the stable coexistence of two species on two limiting resources. In order to investigate the stability properties of the equilibrium point, we will also have to consider the relative resource consumption rates of the two species. For each species, we can associate a resource consumption vector, which will be a directed line with origin at the equilibrium point, and slope equal to the ratio of the resource consumption rates.

Figure 3.8 shows the isoclines with respect to two essential resources (I and II) for two species (1 and 2), such that neither species is competitively superior for both resources. In the situation where the consumption vector u belongs to species 1 and vector v to species 2, a slight increase in species 1 biomass would result in a proportionally larger decrease in the equilibrium level of resource I than resource II. The increase in species 1 biomass would therefore decrease the growth rate of species 2 more than it would for species 1, thus giving species 1 a growth advantage that would lead to further increase in species 1 biomass. The resulting positive feedback loop would eventually lead to the exclusion of species 2. A similar positive feedback cascade leading to the exclusion of species 1 could be initiated by a slight increase in species 2 biomass.

In the opposite situation, where the consumption vector u belongs to species 2 and vector v to species 1, a slight increase in species 1 biomass would result in a proportionally larger decrease in the equilibrium level of resource II than resource I. The increase in species 1 biomass would therefore decrease the growth rate of species 1 more than it would for species 2, thus giving species 2 a growth advantage that would prevent species 1 from increasing its biomass any further. The resulting negative feedback loop would eventually restore the equilibrium biomasses of both species after a perturbation.

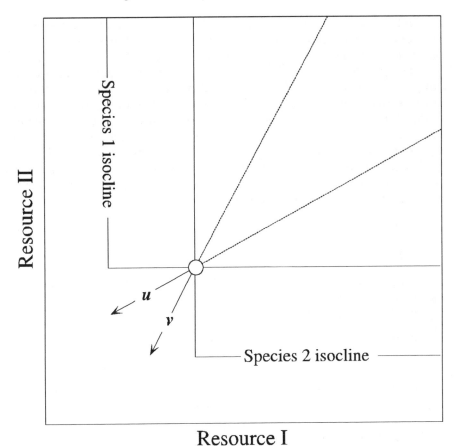

Fig. 3.8. Zero net growth isoclines for two species (*solid lines 1* and *2*) competing for two essential resources (*I* and *II*). *Directed lines* labeled *u* and *v* are different resource consumption vectors

For two species to coexist on two essential resources at a stable equilibrium point, a pair of sufficient conditions would be that (1) neither species is the superior competitor on both resources, and (2) each species consumes proportionally more of the resource on which it is the inferior competitor, at the equilibrium point. When these two conditions are fulfilled, stable coexistence would result from resource supply ratios within the range spanned by the consumption vectors, while competitive exclusion would result for supply ratios outside this range (Tilman 1980). It should be noticed that when both species are assumed to have a constant biomass yield per unit resource, the consumption vectors will be parallel to a straight line from the intersection of the isocline segments to the origin, so that condition (2) will automatically be satisfied whenever condition (1) is. On the other hand, this simplification does not apply if the two species are assumed to have flexible internal stores of the two resources.

A Threshold Model for N- and P-Llimited Phytoplankton Growth. Before we can discuss conditions for coexistence under N- and P-limitation, we must extend the model of P-limited phytoplankton growth to cover N- or P-limited growth according to a threshold model. Assume that we have a pair of phytoplankton species labeled 1 and 2, as in Fig. 3.8, with the ability to form a stable competitive equilibrium for a certain range of N:P supply ratios. For the moment we will also assume that the species have equal maximal growth rates ($\mu_1'' = \mu_2'' = \mu''$; day^{-1}) and that dilution (D; day^{-1}) is the only loss process. If we use the convention that element R denotes either N or P, we can represent the amount of element R in species i as R_i [(μg R) l^{-1}]. The equivalents of the mass-balance equations (2.7), (2.8) for C, N, and P in species i can then be written as

$$\dot{R}_i = v_{R,i}\,C_i - D\,R_i \,, \tag{3.18}$$

$$\dot{C}_i = \left(\mu_i - D\right)C_i \tag{3.19}$$

The specific uptake rate of element R by species i [$v_{R,i}$; (μg R) (mg C)$^{-1}$ day^{-1}] is assumed to follow first-order kinetics as in Section 3.2, but with negligible efflux. If we denote the concentration of dissolved inorganic R by S_R [(μg R) l^{-1}], then the specific uptake rate becomes

$$v_{R,i} = \alpha_{R,i}\,S_R , \tag{3.20}$$

with the uptake affinity [$\alpha_{R,i}$; l (mg C)$^{-1}$ day^{-1}] as a linear decreasing function of the R:C ratio in species i [$Q_{R,i}=R/C_i$; (μg R) (mg C)$^{-1}$], as in Eq. (3.5):

$$\alpha_{R,i} = \alpha_{R,i}' \frac{Q_{R,i}'' - Q_{R,i}}{Q_{R,i}'' - Q_{R,i}'} . \tag{3.21}$$

In Eq. (3.21), $\alpha_{R,i}'$ is the maximal uptake affinity for element R by species i, while $Q_{R,i}''$ and $Q_{R,i}'$ are the maximal and minimal R:C ratios in species i.

The growth rate of species i is controlled by the cellular C:N:P ratios according to a threshold model, such that [μ_i = Min($\mu_{N,i},\mu_{P,i}$), where $\mu_{R,i}$ (day^{-1}) is the growth rate given by the Droop model for element R:

$$\mu_{R,i} = \mu_{R,i}' \left(1 - Q_{R,i}'/Q_{R,i}\right). \tag{3.22}$$

The asymptotic growth rate of the Droop model ($\mu_{R,i}'$) is related to the storage capacity for element R ($Q_{R,i}'' / Q_{R,i}''$), as in Eq. (3.3):

$$\mu_{R,i}' = \mu'' \left(1 - Q_{R,i}'/Q_{R,i}''\right)^{-1} . \tag{3.23}$$

The model is completed by the mass-balances for the dissolved inorganic concentrations of N and P. If R_L [(μg R) l^{-1}] is the input concentration of element R, this mass-balance can be written, in analogy with Eq. (2.13), as

$$\dot{S}_R = D\left(R_L - S_R\right) - \sum_i v_{R,i}\,C_i . \tag{3.24}$$

Constructing a Model Species Pair Capable of Coexistence: In Appendix A1 it is shown that the inequalities (A1.1), (A1.2), (A1.8) define the three sufficient conditions for a stable coexistence of a pair of phytoplankton species under certain N:P supply ratios. By examining Eqs. (A1.1), (A1.2), (A1.8), we find that they can all be expressed in terms of ratios of uptake affinities among the two species. This means that Eqs. (A1.1), (A1.2), (A1.8) confine a triangular feasible region in the plane of relative P and N uptake affinities ($\alpha'_{P,2}/\alpha'_{P,1}$ and $\alpha'_{N,2}/\alpha'_{N,1}$), as shown in Fig. 3.9.

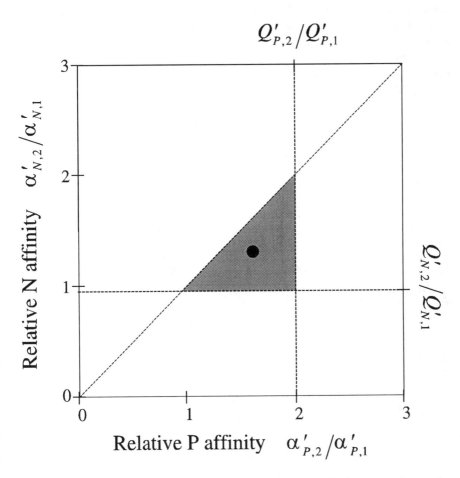

Fig. 3.9. Feasible region for coexistence (*solid triangle*) in terms of relative P and N uptake affinities, as defined by the inequalities Eqs. (A1.1)-(A1.2), (A1.8) (*dashed lines*). *Solid circle* corresponds to the parameter combination given by Table 3.4

If we exchange the roles of species 1 and 2 (such that species 1 becomes the superior competitor for N and species 2 for P), the corresponding feasible region for coexistence would be a reflection across the diagonal line of the region in Fig. 3.9. The feasible region vanishes whenever $Q'_{N,2}/Q'_{N,1} \geq Q'_{P,2}/Q'_{P,1}$ – thus, a necessary condition for coexistence must be that $Q'_{N,2}/Q'_{N,1} < Q'_{P,2}/Q'_{P,1}$, or $Q'_{N,2}/Q'_{P,2} < Q'_{N,1}/Q'_{P,1}$, which means that species 1 must have the highest "optimum" N:P ratio (sensu Rhee and Gotham (1980)).

Figure 3.10 shows the cumulative frequency distribution of a collection of published "optimum" N:P ratios (Table A10.6); that is, the ratio between subsistence quotas for N and P (Q'_N/Q'_P). The observations in Fig. 3.10 are approximately lognormally distributed with a range from 3.2 to 17.6 (μg N) (μg P)$^{-1}$ and a median of 9.0 (μg N) (μg P)$^{-1}$. Inspection of Table A10.6 reveals no distinct pattern among the major taxonomic groups, and shows a rather puzzling variability among observations from closely related species, especially within the genus *Microcystis*.

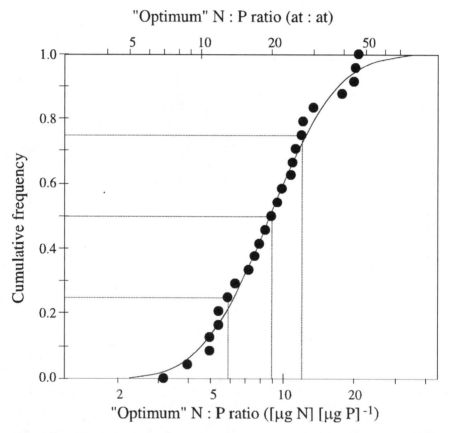

Fig. 3.10. Cumulative frequency distribution of "optimum" N:P ratios in different species of phytoplankton algae (data from Table A10.6; $n = 25$). *Solid line* Fitted lognormal distribution; *broken lines* median and upper and lower quartiles

By the same kind of reasoning as in Section 3.4, we can assign the lower and upper quartiles of the distribution in Fig. 3.10 as low and high parameter values. Thus, we will assume that $Q'_{N,1}/Q'_{P,} = 12.2$ (µg N) (µg P)$^{-1}$ and that $Q'_{N,2}/Q'_{P,2} = 5.9$ (µg N) (µg P)$^{-1}$. Since N is a much larger cell constituent than P, it is reasonable to assume that much of the variability in "optimum" N:P ratios is due to variability in the subsistence quota for P. If we assign the low and high minimum P:C ratios given in Table 3.1 to species 1 and 2 [that is, $Q'_{P,1} = 2.5$ (µg P) (mg C)$^{-1}$ and $Q'_{P,2} = 5.2$ (µg P) (mg C)$^{-1}$], the corresponding subsistence quotas for N become $Q'_{N,1} = 30.5$ (µg N) (mg C)$^{-1}$ and $Q'_{N,2} = 30.7$ (µg N) (mg C)$^{-1}$. If we assume identical storage capacities for P in both species ($Q''_{P,1}/Q'_{P,1} = Q''_{P,2}/Q'_{P,2} = 7.5$; typical parameter value in Table 3.1), the maximum P:C ratios become $Q''_{P,1} = 18.8$ (µg P) (mg C)$^{-1}$ and $Q''_{P,2} = 39.0$ (µg P) (mg C)$^{-1}$. Plankton algae are generally found to have lower storage capacities for N than for P (Turpin 1988), so that we would expect that $Q''_{P,i}/Q'_{P,i} > Q''_{N,i}/Q'_{N,i}$. If we assume $Q''_{N,1}/Q'_{N,1} = Q''_{N,2}/Q'_{N,2} = 4$, as indicated by Goldman and McCarthy (1978), the maximum N:C ratios become $Q''_{N,1} = 121.9$ (µg N) (mg C)$^{-1}$ and $Q''_{N,2} = 122.8$ (µg N) (mg C)$^{-1}$. This set of model parameters gives atomic C:N:P ratios at nutrient-saturated growth equal to 137.4:14.4:1 for species 1, and 66.2:7.0:1 for species 2.

With this combination of subsistence quotas for N and P in the two species, the relative uptake affinities for N and P must be located within the triangular region in Fig. 3.9 in order to have coexistence between the two species at certain N:P supply ratios. If we assume that species 1 has maximum uptake affinities for both N and P equal to the typical parameter value given in Table 3.1 [$\alpha'_{P,1} = \alpha'_{N,1} = 6.5$ l (mg C)$^{-1}$ day^{-1}], then choosing $\alpha'_{P,2} = 10.5$ l (mg C)$^{-1}$ day^{-1} and $\alpha'_{N,2} = 8.5$ l (mg C)$^{-1}$ day^{-1} yields a parameter combination close to the center of the feasible region for coexistence in Fig. 3.9. The resulting set of model parameters are summarized in Table 3.4.

Table 3.4. N and P uptake and utilization parameters in the species pair considered in Fig. 6.13, as well as the corresponding Monod half-saturation parameters [given by Eq. (3.11)]

	Parameter	Species 1	Species 2	Unit
Q'_P	Subsistence P quota	2.5	5.2	(µg P) (mg C)$^{-1}$
Q''_P	Maximal P quota	18.8	39.0	(µg P) (mg C)$^{-1}$
α'_P	Maximal P uptake affinity	6.5	10.5	l (mg C)$^{-1}$ day^{-1}
K'_P	Monod parameter	0.46	0.59	(µg P) l^{-1}
Q'_N	Subsistence N quota	30.5	30.7	(µg N) (mg C)$^{-1}$
Q''_N	Maximal N quota	121.9	122.8	(µg N) (mg C)$^{-1}$
α'_N	Maximal N uptake affinity	6.5	8.5	l (mg C)$^{-1}$ day^{-1}
K'_P	Monod parameter	5.63	4.33	(µg N) l^{-1}

Algal Competition in a Gradient of N:P Supply Ratios. The eight differential equations given by Eqs. (3.18), (3.19), (3.24), together with process rates and parameters given by equations (3.20) - (3.23) and Table 3.4, constitute a chemostat-type model of two species competing for N and P. At a fixed dilution rate (D), a gradient of N:P supply ratios can be simulated by running the model under varying input concentrations of inorganic N and P (N_L and P_L). Since we must have $R_L > S^*_{R,i}$ for R = N, P, in order to have a non-zero biomass of species i at a given dilution rate, input concentrations were chosen such that $R_L > \mathrm{Max}(S^*_{R,1}, S^*_{R,2})$. Using the numerical methods described in Appendix A9, the model was run to equilibrium for different combinations of dilution rates and input concentrations. Model runs were initiated by simulating the inoculation of nutrient-sufficient cells from both species (that is, $Q_{R,i} = Q''_{R,i}$ for i = 1, 2 and R = N, P) into a sterile system with $S_P = P_L$ and $S_N = N_L$ (trials with other choices of initial conditions did not affect the equilibrium solution). Fig. 3.11 shows the simulation results expressed as equilibrium biomass fractions of species 1 for different combinations of N:P supply ratios and dilution rates.

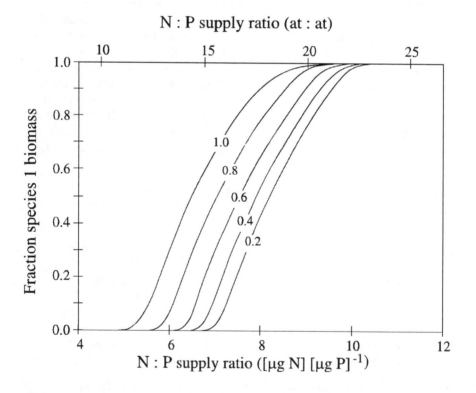

Fig. 3.11. Coexistence in a gradient of N:P supply ratios at different dilution rates. *Solid curves* Equilibrium fraction species 1 of total biomass at a given dilution rate (*curve label*)

Figure 3.11 shows that species 2 (which was designed to be the superior competitor for N) dominates at low N:P supply ratios and is gradually displaced by species 1 (the superior competitor for P) with increasing supply of N relative to P. As pointed out by Turpin (1988), the range of N:P supply ratios supporting coexistence is not independent of the dilution rate; on the contrary, it is shifted toward the lower end of the N:P gradient by increasing growth rate. This downward displacement of the N:P interval of coexistence is probably caused by the assumption of less storage capacity for N than P, which makes the equilibrium cellular N : P ratio at equal N and P limitation ($Q^*_{N,i}/Q^*_{P,i}$) decrease with increasing growth rate for both species.

The two parameter sets given in Table 3.4 are carefully designed to satisfy the conditions for coexistence given by Eqs. (A1.1), (A1.2), (A1.8). If, on the other hand, the two competing species were such that their relative uptake affinities were not within the triangular region in Fig. 3.9, but still gave the same relative isocline positions as in Fig. 3.8, we would have found no region of coexistence. Instead, we would have found a sharp transition from species 2 to species 1 at a N:P supply ratio equal to $S^*_{N,1}/S^*_{P,2}$. If we maintain that $\mu''_1 = \mu''_2$, we would have $S^*_{N,1}/S^*_{P,2} = K^*_{N,1}/K^*_{P,2}$, and therefore that this critical N:P supply ratio is independent of the dilution rate.

3.8 Summary and Conclusions

In the course of this chapter we have developed a simple model of P-limited growth in phytoplankton populations. The model is able to represent a suite of observed phenomena related to algal growth and nutrient uptake including: flexible nutrient utilization in response to nutrient supply rate, fast bidirectional nutrient exchange processes between cells and water, enhanced nutrient uptake capacity with increasing nutrient limitation, and threshold effects in the growth response to external nutrient concentration. The present model is a member of the same family of nonequilibrium extensions to the Monod model, but is able to account for a larger number of phenomena with the same number of parameters as in models of Droop (1983), Burmaster (1979), and Morel (1987). It is shown that this model framework can also be applied to other food web components that utilize dissolved inorganic nutrients (like bacteria), and that it can be extended to represent situations where more than one nutrient may be limiting (like N and P).

The model contains five key parameters that are considered fundamental to the processes of phytoplankton growth and nutrient utilization under phosphorus limitation: maximum growth rate (μ'), minimum and maximum cell quota (Q' and Q''), maximum nutrient uptake affinity (α'), and threshold nutrient concentration for positive net uptake (S'). As very few phytoplankton species have been characterized by the full set of parameters, an attempt has been made to locate realistic parameter values for the typical phytoplankton species in terms of the medians of parameter distributions assembled from an (admittedly incomplete) collection of literature data (Tables A10.1-A10.4). The generalized phytoplankton model developed in this chapter will form the basis for elaborating a more complete representation of pelagic nutrient cycles, where also the role of zooplankton as consumers of autotrophic biomass and regenerators of nutrients is taken into consideration.

3.9 Symbols, Definitions, and Units

Symbol	Definition	Dimension
α	Specific P uptake affinity	$l \, (mg \, C)^{-1} \, day^{-1}$
α'	Maximum specific P uptake affinity	$l \, (mg \, C)^{-1} \, day^{-1}$
$\alpha'_{N,i}$	Maximal N uptake affinity in species i	$l \, (mg \, C)^{-1} \, day^{-1}$
$\alpha'_{P,i}$	Maximal P uptake affinity in species i	$l \, (mg \, C)^{-1} \, day^{-1}$
$\alpha_{R,i}$	Specific uptake affinity for element R in species i	$l \, (mg \, C)^{-1} \, day^{-1}$
$\alpha'_{R,i}$	Maximal uptake affinity for element R in species i	$l \, (mg \, C)^{-1} \, day^{-1}$
$\alpha^*_{R,i}$	Equilibrium uptake affinity for element R in species i	$l \, (mg \, C)^{-1} \, day^{-1}$
φ	Chlorophyll:carbon ratio	$(\mu g \, chla) \, (mg \, C)^{-1}$
φ'	Chla:C ratio at maximum growth rate	$(\mu g \, chla) \, (mg \, C)^{-1}$
K'	Monod half-saturation parameter	$(\mu g \, P) \, l^{-1}$
$K'_{P,i}$	Monod parameter for P-limited growthin species i	$(\mu g \, P) \, l^{-1}$
$K'_{N,i}$	Monod parameter for N-limited growth in species i	$(\mu g \, N) \, l^{-1}$
$K'_{R,i}$	Monod parameter for R-limited growth in species i	$(\mu g \, R) \, l^{-1}$
K_m	Michaelis-Menten half-saturation parameter	$(\mu g \, P) \, l^{-1}$
μ	Specific growth rate	day^{-1}
μ_0	Maintenance growth rate	day^{-1}
μ'	Asymptotic growth rate at infinite P quota	day^{-1}
μ''	Maximum growth rate	day^{-1}
$\mu_{P,i}$	P-limited growth rate in species i	day^{-1}
$\mu_{N,i}$	N-limited growth rate in species i	day^{-1}
$\mu_{R,i}$	Element R-limited growth rate in species i	day^{-1}
$\mu'_{R,i}$	Asymptotic growth rate for R-limited growth in species i	day^{-1}
Q	Cellular P quota	$(\mu g \, P) \, (mg \, C)^{-1}$

Symbol	Definition	Dimension
Q'	Minimum P quota	$(\mu g\ P)\ (mg\ C)^{-1}$
Q''	Maximum P quota	$(\mu g\ P)\ (mg\ C)^{-1}$
$Q'_{P,i}$	Minimum P quota in species i	$(\mu g\ P)\ (mg\ C)^{-1}$
$Q''_{P,i}$	Maximum P quota in species i	$(\mu g\ P)\ (mg\ C)^{-1}$
$Q'_{N,i}$	Minimum N quota in species i	$(\mu g\ N)\ (mg\ C)^{-1}$
$Q''_{N,i}$	Maximum N quota in species i	$(\mu g\ N)\ (mg\ C)^{-1}$
$Q_{R,i}$	Element R quota in species i	$(\mu g\ R)\ (mg\ C)^{-1}$
$Q'_{R,i}$	Minimal element R quota in species i	$(\mu g\ R)\ (mg\ C)^{-1}$
$Q''_{R,i}$	Maximal element R quota in species i	$(\mu g\ R)\ (mg\ C)^{-1}$
$Q^{*}_{R,i}$	Equilibrium R : C ratio in species i	$(\mu g\ R)\ (mg\ C)^{-1}$
S	Dissolved inorganic P concentration	$(\mu g\ P)\ l^{-1}$
S'	Inorganic P concentration at zero net uptake	$(\mu g\ P)\ l^{-1}$
S_P	Dissolved inorganic P concentration	$(\mu g\ P)\ l^{-1}$
$S^{*}_{P,i}$	Equilibrium conc. for P-limited growth in species i	$(\mu g\ P)\ l^{-1}$
S_N	Dissolved inorganic N concentration	$(\mu g\ N)\ l^{-1}$
$S^{*}_{P,i}$	Equilibrium conc. for N-limited growth in species i	$(\mu g\ N)\ l^{-1}$
S_R	Concentration of element R in dissolved inorganic form	$(\mu g\ R)\ l^{-1}$
$S^{*}_{N,i}$	Equilibrium conc. for R-limited growth in species i	$(\mu g\ R)\ l^{-1}$
v	Specific net P uptake rate	$(\mu g\ P)\ (mg\ C)^{-1}\ day^{-1}$
v_i	Specific gross P influx rate	$(\mu g\ P)\ (mg\ C)^{-1}\ day^{-1}$
v_e	Specific gross P efflux rate	$(\mu g\ P)\ (mg\ C)^{-1}\ day^{-1}$
v_m	Saturated net P uptake rate	$(\mu g\ P)\ (mg\ C)^{-1}\ day^{-1}$
v'_m	Maximum saturated net P uptake rate	$(\mu g\ P)\ (mg\ C)^{-1}\ day^{-1}$
v''_m	Minimum saturated net P uptake rate	$(\mu g\ P)\ (mg\ C)^{-1}\ day^{-1}$
$v_{R,i}$	Specific uptake rate of element R by species i	$(\mu g\ R)\ (mg\ C)^{-1}\ day^{-1}$
Y	Biomass growth yield	$(mg\ C)\ (\mu g\ P)^{-1}$
Y'	Maximum growth yield	$(mg\ C)\ (\mu g\ P)^{-1}$

4 Herbivores and Algae: Food Utilization, Growth and Reproduction in Generalist Filter Feeders

The pelagic zone of the ocean or a large lake represents an essentially limitless food supply for the individual zooplanktonic animal, but, at the same time, this source of energy may be so thinly dispersed that the problem of getting enough to eat is still formidable.

Conover (1968).

In a biogeographic perspective, almost all lakes can be considered ephemeral environments. The constraints of using resting stages for dispersal and risk avoidance in such an environment is probably a major reason why only the Rotifera and a few orders of Arthropoda (mainly Cladocera, Calanoida, and Cyclopoida) have been successful in exploiting the pelagic zone of lakes, while marine zooplankton also have representatives from at least six other phyla (Lehman 1988). When compared to the multitude of species commonly found in the phytoplankton of fresh waters, the zooplankton of lakes generally appear to be species-poor communities (Pennak 1957).

Freshwater zooplankton community structure is closely linked to the selective feeding activity of vertebrate predators (Brooks and Dodson 1965). Prey selection in planktivorous fish is generally determined by prey visibility, such that large herbivore species like *Daphnia* will be selectively removed when planktivore biomass is high. On the other hand, large filter-feeders like *Daphnia* seem to be competitively superior and tend to dominate the zooplankton community in the absence of strong vertebrate predation pressure. The key competitive advantage of *Daphnia* has not been unambiguously identified, though several alternative explanations have been proposed (see, for example, the recent review by DeMott 1989). The presence or absence of large *Daphnia* species has also been shown to have greater impact on the phytoplankton community than total zooplankton abundance (Pace 1984; McQueen et al. 1986), emphasizing the key role of the genus *Daphnia* in the pelagic food web.

de Bernardi and Peters (1987) have documented the ubiquitous use of *Daphnia* species as model organisms in ecological research on predation, energy flow, population regulation, competition, and evolution. It seems likely that more research effort has been devoted to *Daphnia* than to any other genus of freshwater invertebrates. The abundance of *Daphnia* publications can be defended partly by the ecological importance in plankton communities, and partly by the conformity between many aspects of *Daphnia* biology and general body-size relationships describing "typical" animals (de Bernardi and Peters 1987).

Within the framework of the simple food chain outlined in Section 2.5, *Daphnia* seems a natural choice for a model organism when representing the processes of grazing, growth and remineralization in the pelagic phosphorus cycle. This chapter will therefore mainly be devoted to the formulation and parametrization of a model of nutrition and growth in *Daphnia* populations, but the main principles should be directly applicable to other cladoceran species, and also to herbivorous rotifers and copepods, with some minor modifications.

4.1 Characteristics of *Daphnia* Biology

Generation Cycle. In *Daphnia* and other cladocerans the basic generation cycle consists of consecutive runs of parthenogenic generations interrupted by periods of diapause, initiated by the production of resting eggs. According to Zaffagnini (1987), the ability of *Daphnia* to reproduce both parthenogenetically and sexually was already well known in the 1850s. Although the presence of males is often coincident with production of diapausing eggs in *Daphnia*, male fertilization does not seem to be necessary to produce resting eggs in some populations (Stross 1987).

Stross (1987) indicates that the induction of diapause in *Daphnia* is probably controlled by a combination of photoperiod and food conditions. Diapause starts with embryo development being interrupted near the gastrula stage, after which the two resting eggs are expelled into an ephippium. The embryo can resume development after a prescribed interval of dormancy, and ephippia seem to remain viable for several years. De Stasio (1989) has suggested that copepod resting eggs stored in the sediment constitute an egg bank analogous to the seed banks of higher plants, and similar arguments can be made for *Daphnia* ephippia stored in the sediments. Threlkeld (1987) has argued that ephippia are relatively unimportant in permanent lakes with *Daphnia* populations overwintering in the water column, although it seems likely that the egg bank can at least be a source for *Daphnia* to recolonize a lake after accidental extinction.

The Fertility Cycle of Parthenogenetic Reproduction. Under normal conditions, parthenogenic eggs are released from the oviducts into the brood chamber following the maternal molt, while the new female carapace is still soft and elastic (Zaffagnini 1987). Most cladoceran genera, including *Daphnia*, do not secrete any nutritional fluid into the brood pouch, so that embryonic development is exclusively at the expense of resources deposited with the egg in the form of yolk and lipid droplets (Green 1956). Much recent work on *Daphnia* energetics has been focused on triacylglycerol lipids (e.g., Goulden et al. 1984), although it seems certain that *Daphnia* storage products also include many other macromolecules like glycogen and lipoproteins (Zaffagnini 1987; Elendt 1989).

Crustaceans are unique in the sense that yolk material can be synthesized outside the ovaries (extraovarian vitellogenesis; Meusy 1980), and combined with the rest of the egg at a late stage in oogenesis. Ultrastructural studies by Zaffagnini (1987) indicate that the yolk of parthenogenetic eggs in *Daphnia* is entirely of extraovarian origin, probably synthesized by the fat cells surrounding the gut. According to Goulden et al. (1984), the fate of the storage products intended for yolk synthesis seems to be undetermined until the later stages of egg development; under acute starvation they might be metabolized by the adult herself instead of being transferred to the eggs. The fraction of the maternally supplied yolk that remains after embryonic development is used to support growth in the early juvenile instars (Tessier and Goulden 1982).

Individual Life History Events. *Daphnia* egg development is phased to the duration of the intermolt interval (Banta 1939), so that the neonates swim away from the brood pouch shortly before the next molt. The intermolt duration seems to be mainly determined by the temperature and largely unaffected by other environmental factors (Bottrell 1975; Lynch 1989). Unlike many other crustaceans, *Daphnia* development into adulthood is direct without any metamorphosis; neonates are in a way miniature immature adults.

It seems likely that maturation in cladocerans is determined by reaching a critical body size (Lynch 1980a, 1989; Perrin 1989), while the age at first maternity can be prolonged severalfold under low food conditions. In *Daphnia* and several other cladocerans, body growth is continued after reaching the size of maturity (Lynch 1980a). Changes in carapace size are limited to the first few minutes after molting (Green 1956), after which the carapace length stays constant throughout the rest of the instar while the body mass will be continuously changing. Parthenogenetic reproduction in *Daphnia* is iteroparous, with a female being able to produce up to 25 clutches in her lifetime under favorable conditions.

Nutrition and Energy Allocation. The relationships between the major structural components in *Daphnia* are illustrated in Fig. 4.1. Ingested food is digested and absorbed by the cells of the lumen in the midgut, while the unassimilated fraction is egested through the hindgut (Peters 1987). Assimilate is transferred to the hemolymph, which serves as the transport medium to other parts of the body. Tracer studies have revealed that the metabolism of carbon and phosphorus in *Daphnia* can both be represented by two compartments; a large compartment with slow turnover, and a small one (1-6% of the total) with fast turnover (<3 h) (summarized in Table 3 of Lampert and Gabriel 1984). It is reasonable to identify the fast compartment with the hemolymph, and to assume that the hemolymph will be in equilibrium with the remaining structural parts of the animal (the slow compartment) on a time scale of days.

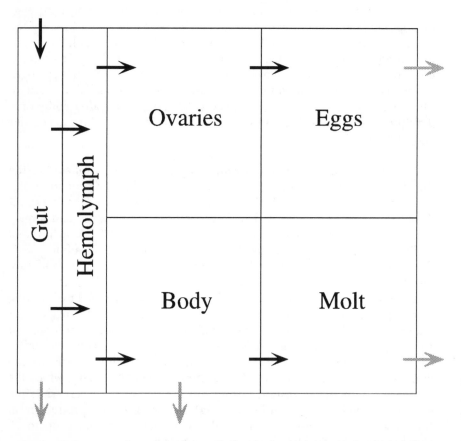

Fig. 4.1. Main structural compartments and allocation pathways in *Daphnia*. *Black arrows* denote compartemental transfers of assimilated food while *gray arrows* denote material losses from the body (see text)

Assimilated food has to support three main processes in growing animals: the generation of new body mass (somatic growth), the production of offspring (reproductive growth), and the requirements of maintenance. In *Daphnia*, maintenance includes both the costs of basal, cellular metabolism to support body functions and the cost of regenerating lost material like exuviae. It is reasonable to assume that maintenance takes priority over the other processes, so that both somatic growth and reproduction are halted when food assimilation is inadequate to cover maintenance needs. When assimilation is sufficient to support growth, experimental evidence suggests that in *Daphnia* the relative allocation into somatic and reproductive growth is dependent only on body size and not on the food supply (Lynch 1989).

4.2 The Allocation of Net Assimilate

Nutrition can be defined as the balance of physiological processes enabling an animal to survive, grow, and reproduce (Conover 1968). As food carbon will be the energy source of an animal, we can expect the energy budget of an animal to be closely related to its carbon mass balance. Under the assumption of a food source matching all the requirements of the animal, the main processes involved in animal nutrition can therefore be represented by the carbon mass balance, where ingested food carbon is either used for growth and reproduction, respired, or egested (e.g., Richman 1958). If we represent carbon egestion as a fraction $1 - \varepsilon$ of carbon ingestion, where ε is the dimensionless food assimilation efficiency, then the basic carbon or energy balance can be written in terms of body-carbon-specific rates as

$$I = g + r + (1 - \varepsilon) I, \qquad (4.1)$$

with I being the specific ingestion rate (day^{-1}), r the specific respiration rate (day^{-1}), and g the specific net assimilation rate (day^{-1}). Collecting the ingestion terms and rearranging gives

$$g = \varepsilon I - r, \qquad (4.2)$$

stating simply that net assimilation is the difference between gross assimilation (εI) and respiration.

Modeling Strategy. Equation (4.2) is a natural starting point for physiological model construction, making growth the output of more or less detailed submodels describing food ingestion, assimilation, and respiration (e.g., Paloheimo et al. 1982; Kooijman 1986; Gurney et al. 1990; McCauley et al. 1990). Lehman (1988) pointed out some pitfalls that might be buried in this approach: if, for example, assimilation and respiration rates have a lognormal distribution (which implies that rates are non-negative, as they should be), the compound distribution of g resulting from Eq. (4.2) will not be lognormal, and will contain both negative and positive values. In other words, taking the difference of lognormal variables tends to amplify the uncertainty in the individual variables to an extent that the result can be nearly meaningless.

Another, and perhaps more serious objection to describing growth from submodels of assimilation and respiration comes from a study by Lynch et al. (1986), who compared *Daphnia* net growth estimates based on two essentially different methods: long-term lifetable studies and short-term physiological measurements. The disparity in experimental time scales apparently created some striking inconsistencies between the two methods; while correspondence was good for juvenile *Daphnia*, life tables gave constant production [as (μg C) ind.$^{-1}$ day^{-1}] in adults, while tracer studies predicted production proportional to animal size throughout their lifespan (Lynch et al. 1986). Assimilation measurements by tracer methods on the time scale of 1-3 h, as used by Lampert and coworkers (Lampert 1977; Lynch et al. 1986), are based on quite a few assumptions that are necessary to correct for respirational losses of fresh assimilate. As direct growth measured in life table experiments contains fundamentally fewer assumptions than tracer-based methods, it is reasonable that one should put more confidence in life table data under conditions where the two approaches diverge.

The modeling strategy chosen here will therefore be first to formulate a model of the allocation of net assimilate (i.e. assimilation less respiration) into growth and reproduction, using life table data as input. In the next section this model will be expanded to include the overheads of egestion and respiration, so that assimilation will be given as the carbon input necessary to support observed growth and maintenance costs.

Net Assimilate Allocation Model. As the molt cycle appears to be a fundamental rhythm in cladoceran growth and reproduction, the instar duration is a natural time scale for an energy-allocation model. At 20 °C, the intermolt period is 1-3 days in *Daphnia*, with the adult instars being approximately twice as long as the juvenile ones. At this time scale there is probably no reason explicitly to model the dynamics of the fast pool identified with the contents of the hemocoel in Fig. 4.1. If we consider the carapace and the eggs carried in the brood pouch as passive components of an adult *Daphnia*, we need to take into account only two carbon pools to describe growth from one instar to the next; body mass (B) and accumulated ovary material that will be extruded as eggs at next molt (E) (both as μg C):

$$\dot{B} = (1 - R)g\,B \qquad\qquad\qquad (4.3)$$

$$\dot{E} = R\,g\,B, \qquad\qquad\qquad\qquad (4.4)$$

with initial conditions $B(0) = B_i$ and $E(0) = 0$ at the start of instar i. At the end of instar i (after the instar duration D_i has elapsed), B equals the body mass at the start of the next instar $(i + 1)$ plus the mass of the exuvium that will be lost at the transition to the new instar (M_i), that is, $B(D_i) = B_{i+1} + M_i$, while E equals the accumulated egg production during instar i; $E(D_i) = E_i$. Both the dimensionless fraction of assimilate allocated to reproduction (R) and the specific net assimilation rate (g) are assumed to be functions of body mass.

The Growth History of Daphnia pulex. As body mass changes continuously within each instar while body length changes only at instar transitions, body weight estimated from body length will be valid only with reference to a specific point in the molt cycle, most commonly at the entry of new instar (Lynch 1989). Carapace length at a given instar i $(L_i;$ mm) and body weight at the entry of the same instar $(B_i;$ µg C) in *Daphnia* are usually found to be related by an allometric, or power function which can be written as:

$$B_i = B_{ref} \left(\frac{L_i}{L_{ref}} \right)^{\beta_B} . \qquad (4.5)$$

If the reference length L_{ref} is 1 mm, then B_{ref} is the body mass of 1-mm-long animals, and β_B is the allometric exponent. By neglecting the reference dimension L_{ref} in power functions like this, as some workers do, the coefficient B_{ref} ends up having a strange and awkward dimension like (µg C) mm$^{-\beta}$, instead of µg C.

Lynch (1989) found that a single allometric relationship could be used to convert length measurements into body masses of freshly molted *Daphnia pulex* individuals raised under widely differing food conditions. If we assume a carbon to dry weight ratio of 0.46 (Andersen and Hessen 1991), the parameters of allometric relation [Eq. (4.5)) for *Daphnia pulex* as given by Lynch (1989) will be $B_{ref} = 3.0$ µg C and $\beta_B = 2.6$.

The size at first maternity is identified as an important life history characteristic in cladocerans (Lynch 1980a). As the first clutch will be visible in the instar after the first reproductive investment, and as initial reproductive investment is small (Fig. 4.3), the size of primiparous females (B'') will be larger than the size of first reproductive investment (B'). Recalculating the lengths of primiparous *Daphnia pulex* females reported by Richman (1958), Paloheimo et al. (1982), Taylor (1985), and Lynch et al. (1986) to carbon unit gives an average size at first maternity $B'' = 13.6$ µg C. Although Lynch (1989) states that some *Daphnia pulex* individuals can reach an ultimate size of 3.5 mm, the maximal sizes reported in his data are slightly above 3.1 mm, corresponding to a maximal body mass $B''' = 60$ µg C.

Lynch et al. (1986) showed that the body mass lost by molting at end of a given instar i $(M_i;$ µg C) in *Daphnia pulex* is also an allometric function [Eq. (4.6)] of carapace length L_i; with reference molt loss $M_{ref} = 0.24$ µg C and exponent $\beta_M = 3.1$.

$$M_i = M_{ref} \left(\frac{L_i}{L_{ref}} \right)^{\beta_M} . \tag{4.6}$$

Lynch (1989) found a significant variation in egg size with food conditions, but the effect was so small (<10%) that it can in practice be assumed that egg carbon content in *Daphnia pulex* is a constant $B_0 = 1.1$ μg C, which is not significantly different from the neonate body mass (B_1; μg C) reported by the same author. Goulden et al. (1984) also found only a slight and insignificant difference in dry weight between eggs and neonates in *D. galeata* and *D. magna*. The small mass decrease during egg development is supported by the results of Bohrer and Lampert (1988), who found very low respiration rates in *D. magna* eggs (< 20% of the rates found in juveniles of the same species). It can therefore be assumed that the initial body mass in the first instar is equal to the mass of a fresh egg ($B_0 = B_1$).

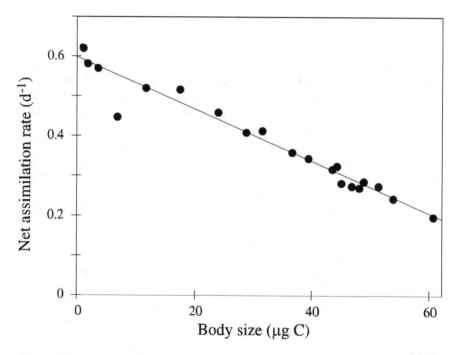

Fig. 4.2. Specific net assimilation rate [estimated from Eq. (A2.7)] as function of body size in *Daphnia pulex* [pooled data from food levels > 1 (mg C) l[-1] from Lynch 1989]. *Solid line* is given by the model given by Eq. (4.7)

Lynch (1989) reported a comprehensive life table study on *Daphnia pulex* reared at nine different food concentrations ranging from 3.1 (mg C) l^{-1} down to 0.015 (mg C) l^{-1}. Animals were cultured individually in 40 ml volumes with food replenishment every other day. Results from the three highest food concentrations [>1 (mg C) l^{-1}] were very similar, indicating that these represented animals growing at their inherent capacity, without any food limitation. This conclusion is also supported by the results of Lynch et al. (1986) which were unchanged by a fivefold increase in culture volume, giving a food supply equivalent to 7.5 (mg C) l^{-1}.

The carbon budget terms (B_i, E_i, and M_i) for each instar i can be calculated from combining the instar-specific carapace lengths, clutch sizes, and duration times reported in Lynch (1989) with the allometric relationships Eqs. (4.5), (4.6). From the carbon budget terms the net assimilation rate g_i and the fraction of net assimilate allocated to reproduction R_i can be estimated for every instar as shown in Appendix A2. Figure 4.2 shows the specific net assimilation rate [g; estimated from Eq. (A2.7)] as function of body mass (B). The decrease in g with size is clearly very close to linear, and can be represented by a straight line as

$$g = g_1 \frac{B_\infty - B}{B_\infty - B_1},\tag{4.7}$$

with $g_1 = 0.59$ day-1, and $B_\infty = 93$ µg C. g_1 can be interpreted as the net assimilation rate of a newly hatched neonate ($B = B_1$), while B_∞ is the asymptotic body size above which no positive growth is possible ($g < 0$). The drop in the net assimilation rate at the last preadult instar is probably a real phenomenon reflecting the extra costs of the transition to adulthood, although this cannot be reproduced in the linear model [Eq. (4.7)].

Due to the wide body size span in *Daphnia pulex*, the growth rate over the range of juvenile instars will be close to g_1 (<10% reduction from neonate to primiparous instar). While Eqs. (4.3) and (4.7) together predict a logistic growth curve in juveniles, the large difference between the size at maturity and the asymptote of the logistic curve (B_∞) makes the juvenile body mass development very close to exponential. This is in accordance with the results of Tessier and Goulden (1987), who found that the juvenile development in several species of cladocerans could be well represented by an exponential growth model.

Figure 4.3 shows the fraction of net assimilate allocated to reproduction (R) as a function of body mass; the first investment to reproduction is made at a body size of 7-8 µg C, after which the fraction rises steeply to an asymptotic level around 0.8. Many different functions that could be fitted to this pattern; a simple representation is the truncated inverse quadratic:

$$R = R' \left(1 - \left(\frac{B'}{B}\right)^2\right)_+,\tag{4.8}$$

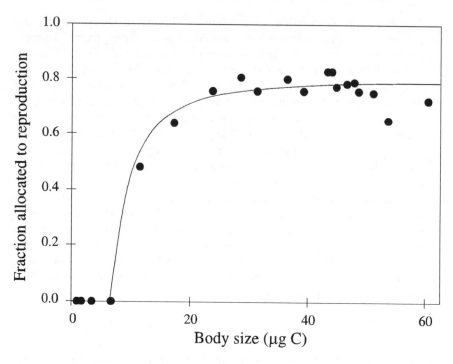

Fig. 4.3. Fraction of net assimilate allocated to reproduction [estimated from Eq. (A2.2)] as a function of body size in *Daphnia pulex* [pooled data from food levels > 1 (mg C) l⁻¹, from Lynch 1989]. *Solid line* is the model given by Eq. (4.8)

where the operator $(\ )_+$ is zero whenever the expression inside the paren-thesis is ≤ 0. With $R' = 0.8$ and $B' = 7.5$ µg C, Eq. (4.8) gives an acceptable fit (Fig. 4.3). B' is then interpreted as the size of first investment to reproduc-tion and R' is the asymptotic level of reproductive investment. The discon-tinuity at B' is an unavoidable consequence of the life history pattern in *Daphnia*. The decrease in reproductive investment in very large animals might be a real phenomenon caused by increasing molt costs, but it is probably not worthwhile to introduce an additional parameter to model this.

Equations (4.3)-(4.6) constitute a model of body growth and reproduc-tion between instar transitions in *Daphnia pulex*. In Fig. 4.4 the differential equations (4.3) and (4.4) were solved numerically (using the methods described in Appendix A9), with initial conditions $B = B_l$ and $E=0$, where B_l is the body mass of a newly hatched neonate (Table 4.1). At the end of each instar, molting and egg laying were simulated as $B \leftarrow B - M$ and $E \leftarrow 0$, where M is the molt weight at the start of the previous instar ($v \leftarrow S$ means that the variable v is overwritten by the result of the expression S).

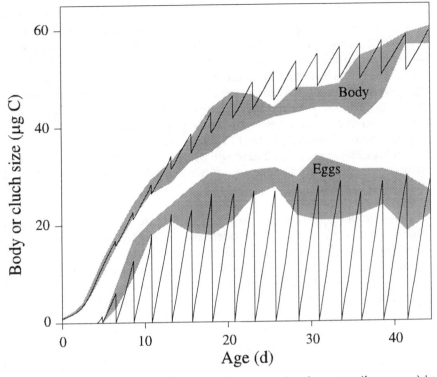

Fig. 4.4. Simulated development of body mass (*upper curve*) and egg mass (*lower curve*) in *Daphnia pulex* growing in surplus food. *Shaded areas* are ranges of body and clutch sizes (recalculated to µg C) for animals grown at food concentrations >1 (mg C) l^{-1}, as reported by Lynch (1989)

The growth pattern predicted by the simulations (Fig. 4.4) reproduces quite nicely the main life-history events in *Daphnia pulex*. The largest discrepancy is found between observed and predicted adult body mass (especially from an age of 25 days and onwards), while correspondence between observed and predicted egg mass is uniformly good. As so much of adult production is devoted to reproduction, total production will be predicted with better accuracy than body mass.

Table 4.1. Body size characteristics in the growth history of *Daphnia pulex*

	Parameter	Value
B_0	Egg size	1.1 µg C
B_1	Neonate size	1.1 µg C
B'	Size at first reproductive investment	7.5 µg C
B''	Size at first maternity (primipara)	13.6 µg C
B'''	Maximum attainable size	60.0 µg C

Fig. 4.4 also points out the importance of molt losses in determining the growth pattern in *Daphnia*. Although the cost of molting might be small compared to total production, it constitutes a major fraction of the growth investment in adults. The setback created by molt losses gives a strong retardation in adult body mass growth rate, which again is responsible for the nearly constant production rate observed in adult *Daphnia* by Lynch et al. (1986). Running the model without molt losses results in major deviations from the observed pattern. Body growth will proceed toward the asymptotic body mass B_∞ in Eq. (4.7), which is > 50% higher than observed. At the same time net assimilation will decrease as a result of Eq. (4.7), such that the egg production at the end of the simulation is reduced to < 5% of the observed level.

A Continuous Approximation to the Growth History of Daphnia. The continuous-discrete nature of the model [Eqs. (4.3), (4.4)] represents a major complication to its incorporation in a full population model. It is therefore desirable to avoid the explicit modeling of instar transitions by approximating the present model with a fully continuous version. Due to the differences in the exponents of the allometric relationships [Eqs. (4.5), (4.6)], molt weight will increase faster with length than body weight, and thus the molt will constitute an increasing fraction of body weight with increasing size. On the other hand, instar duration will also increase with increasing size, giving larger animals more time to build up the new molt. Assuming that these two forces will work together in such a way that the relative cost of molting will be independent of body mass leads to a particularly simple extension to Eq. (4.3), which can be written as

$$\dot{B} = \big((1 - R)\, g - h\big) B, \tag{4.9}$$

where h is the relative cost of molting in units of day^{-1}. In this new model we cannot describe egg mass accumulation explicitly, as in Eq. (4.4), while the instantaneous egg production can still be calculated as $R\, g\, B$, when the solution of Eq. (4.9) is known.

Equation (4.9) predicts that body growth stops ($\dot{B} = 0$) at a maximum body size (B'''; Table 4.1), when all of the growth investment is lost again through molting; that is when $(1- R)\, g = h$. Setting $B = B'''$ and $\dot{B} = 0$ in Eq. (4.9) gives $h = (1 - 0.79) \cdot 0.21$ day^{-1} = 0.044 day^{-1}. When comparing the output of the continuous [Eq. (4.9)] and discontinuous models [Eqs. (4.3), (4.4)], quite good correspondence is seen among the predicted adult body sizes and egg production rates (Fig. 4.5). The single differential equation (4.9) describing somatic growth, together with the equations describing net assimilation [Eq. (4.7)] and allocation to reproduction [Eq. (4.8)] as functions of body size, can be considered a minimal representation of the life-history events leading to growth and reproduction in *Daphnia*.

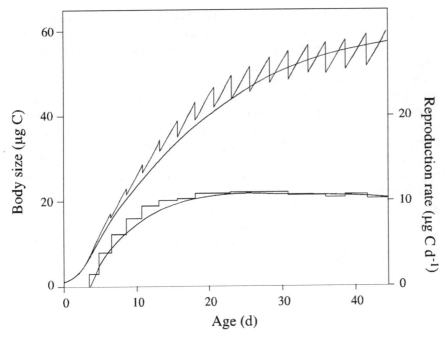

Fig. 4.5. Simulated body mass (*upper curves, left axis*) and egg production rate (*lower curves, right axis*) in *Daphnia pulex* growing on surplus food. Smooth curves are output from the continuous model, given by Eq. (4.9), while discontinuous curves are the same as in Fig. 4.4 with the egg production rate averaged over each instar

4.3 Ingestion, Assimilation, and Overheads on Growth

So far, we have considered only the allocation of net assimilate in *Daphnia* when food supply is sufficient for the animals to grow at their innate capacity. In order to expand the model to cover growth under food limitation, we must take into account the individual terms of the gross carbon budget as described by Eq. (4.2), including the overheads from respiration and the balance between ingestion and egestion.

Respiration. Glazier (1991) found that eggs and early embryos of Daphnia magna had respiration rates that were only a third of those of neonates and adults, and that brooding adults therefore have lower respiration per unit total body mass than juveniles. Both measurements on debrooded females and corrections based on measured respiration rates in developing eggs and embryos indicated that respiration is very nearly isometric with body mass in Daphnia. Glazier (1991) inferred that the often reported negative allometry between respiration and body mass in Daphnia might be an artifact caused by the disproportionately low contribution of egg respiration in brooding adults, and that body-mass-independent specific respiration rates might be common phenomenon in small metazoans, as first suggested by Zeuthen (1953).

Figure 4.6 shows a collection of published data sets on respiration in *Daphnia pulex* as function of body size. Apart from a few very high values from individuals at the size of neonates, there seems to be no strong trend in respiration as a function of body size in this species, although this might to some extent be due to the lack of respiration data from large *D. pulex* (> 25 µg C). If some of the maternally supplied reserves deposited with the egg are carried over to the early juvenile instars, as suggested by Tessier and Goulden (1982), O_2 consumption will overestimate the actual drain of assimilate to maintenance in these stages. Additionally, if some of the maintenance energy in early juvenile instars comes from lipid instead of protein metabolism, the conversion factor from O_2 consumption to carbon loss will be too high.

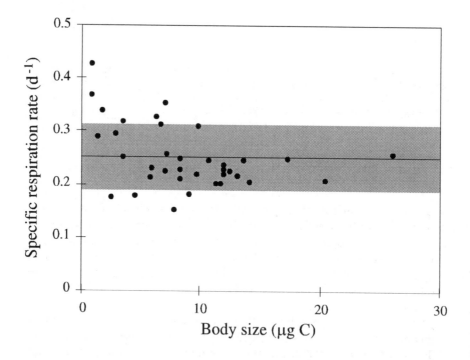

Fig. 4.6. Body mass-specific respiratory carbon loss rates (day[-1]) in *Daphnia pulex* based on data from Richman (1958), Buikema (1972), Kring and O'Brien (1976), Goss and Bunting (1980), and Richman and Dodson (1983). Mean and standard deviation are indicated by *horizontal line* and *shaded area*. Oxygen consumption measurements have been recalculated to body carbon specific loss rates using a conversion factor of 0.38 (µg C) (µg O_2)[-1], which implies protein metabolism and a respiratory quotient of 1.0 (Lampert and Bohrer 1984)

In juveniles, the maintenance costs of basal metabolism (r) and molting (h) will be equivalent in the sense that they will both be taken from the total energy intake of the animal, so that an underestimation in juvenile respiration might to some extent be balanced by the slight overestimation of molting losses in juveniles, implied by the continuous growth model [Eq. (4.9)]. Based on these arguments, it seems that respirational costs are independent of body size, with an average specific, carbon-based loss rate due to respiration of 0.25 day^{-1}, the mean of the data in Fig. 4.6 after excluding three values presumably representative of neonate metabolism. With a constant overhead to respiration, gross assimilation ($g + r$) will decrease with body size, with the same slope as net assimilation (Fig. 4.2).

Assimilation. Although experimental evidence is limited, the results of both Lampert (1977) and Lynch et al. (1986) suggest that assimilation efficiency is independent of body size. Conover (1966) and Lampert (1977) have both disclaimed the occurrence of reduced assimilation efficiency at very high food concentrations (superfluous feeding; Beklemishev 1962) in herbivores, but Lehman (1984) warns that this does not necessarily preclude the possibility of increased digestive efficiency as a result of the prolonged gut passage time at very low food levels (Geller 1975).

On the other hand, most experiments suggest that assimilation efficiency is more related to food quality (in terms of taxonomic and biochemical composition, detritus content, etc.) than to food concentration. For example, Hessen et al. (1990) found that carbon assimilation efficiency in a natural population of *Daphnia longispina* ranged from 70% on algae down to less than 15% on detritus.

For the typical food algae used in laboratory experiments with *Daphnia* (e.g., *Scenedesmus* spp. and *Chlamydomonas* spp.), the assimilation efficiency is probably very high, although the highest values reported (>90%) should be treated with some caution due to the potential biases resulting from inhomogeneous tracer labeling (Lampert 1987; Nielsen and Olsen 1989). Methods avoiding the pitfalls of tracer dynamics indicate a maximum assimilation efficiency around 80% (Olsen et al. 1986a) in *Daphnia* feeding on nutrient-saturated and well-assimilable algae.

Ingestion. If we accept the assumption of size-independent assimilation efficiency and respiration, we can solve the basic carbon balance [Eq. (4.2)] with respect to I and substitute g with the expression (4.7) to obtain the relationship between body size and ingestion rate at saturating food concentrations (I'; day^{-1}):

$$I' = I_1' \frac{B_\infty' - B}{B_\infty' - B_1}, \qquad (4.10)$$

where $I_1' = (g_1 + r)/\varepsilon = 1.05$ day^{-1} and $B_\infty' = B_\infty + (r/g_1)(B_\infty - B_1) = 132$ µg C. In accordance with Eq. (4.7), I_1' can be interpreted as the ingestion rate of neonates, while B_∞' is the extrapolated body size where ingestion becomes zero.

Most published ingestion rate measurements on *Daphnia* have been presented as allometric functions of body length, with an exponent around 2-3 (e.g., Burns and Rigler 1967; Geller 1975; DeMott 1982; Ganf and Shiel 1985). When such measurements are rescaled to body-mass-specific rates as function of body mass, much of the non-linearity disappears, giving functional relationships that might as well be represented by straight lines.

As an example, one can compare the ingestion rates reported by Geller (1975) for *Daphnia pulex* feeding on high and low concentrations of *Scenedesmus* with the predictions from Eq. (4.10). The linear regression slopes of both data sets in Fig. 4.7 are not significantly different ($p > 0.05$) from the slope predicted by Eq. (4.10), while the intercepts are 20-30% lower than the model prediction. The increase in variability at the high food level seems, for unknown reasons, to be a common feature of ingestion rate measurements by tracer methods (Lampert 1987). When taking into account that Eq. (4.10) is based on data completely independent of the data in Fig. 4.7, the overall correspondence must be considered encouraging.

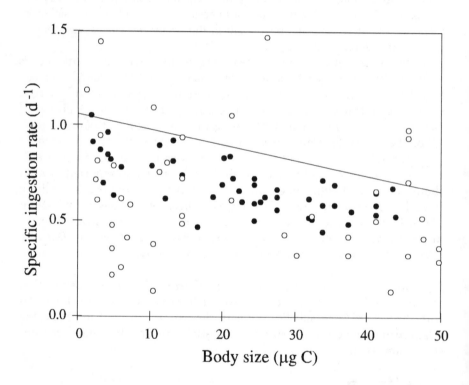

Fig. 4.7. Specific ingestion rate as a function of body size for *Daphnia pulex* feeding on *Scenedesmus*, recalculated from Geller (1975). *Solid symbols* 0.17 (mg C) l^{-1}; *open symbols* 1.52 (mg C) l^{-1}; *straight line* is the ingestion rate predicted from Eq. (4.10)

4.4 Food Supply Regimes and Individual Growth Histories

Rigler (1961) noticed that at low food levels the feeding rate of *Daphnia* seem to be limited by the amounts of water the animal can filter, while at high food levels it is limited by some other factor. As a result, the ingestion rate increases with food concentration until the "incipient limiting level" (McMahon and Rigler 1965) is reached, after which the ingestion rate levels out and becomes independent of the food level. Both reduced filtering rate through lowered appendage beating frequency, and increased rejection rate through more frequent clearing of the food groove by the abdominal claw are probably important mechanisms in determining the functional response in *Daphnia* (Porter et al. 1982).

An optimal foraging model presented by Lehman (1976) predicts that animals should stop feeding at very low food levels in order to save energy. Experimental evidence so far indicates that such a threshold must indeed be very low [at most a few (µg C) l^{-1}; Muck and Lampert 1980; Porter et al. 1982], indicating that the feeding response in *Daphnia* is probably a monotonic increasing function.

The Functional Response. Due to the conspicuous increase in experimental scatter at high food levels in typical *Daphnia* feeding experiments, several different functions can be fitted with almost equal goodness-of-fit (Lampert 1977; Porter et al. 1982). There is some evidence that the transition to saturated ingestion often is more abrupt than would be predicted from the common rectangular hyperbola (the Michaelis-Menten curve of enzyme kinetics), and that a piecewise linear function generally is more appropriate (Condrey and Fuller 1985).

With the notation used so far, the dependence of the specific ingestion rate on food concentration [C; (mg C) l^{-1}] can be written as

$$I = \begin{cases} \dfrac{C}{I'\,C'} & C \le C' \\ I' & C > C' \end{cases} = \frac{I'}{C'}\min(C,C'),$$

(4.11)

where C' is the incipient limiting food level [(mg C) l^{-1}] and I' is the maximum ingestion rate at a given body size[Eq. (4.10)].

The functional response [Eq. (4.11)] contains the implicit assumption that C' is independent of body size. The available experimental data seem to be both limited and ambiguous; McMahon (1965) reported changes in C' with body size, while Lampert (1977) presented feeding curves indicating a constant C' for different size groups of D. pulex grazing on Stichococcus. Most workers have found the incipient limiting food level to be in the range 0.2-0.5 (mg C) l^{-1} for Daphnia feeding on readily ingestible algae (Lampert 1987), although Thompson et al. (1982) have reported values below 0.2 (mg C) l^{-1} for *Daphnia* feeding on natural seston.

Lampert and coworkers have used a flow-through system (Lampert 1976) for growing food-limited zooplankton under controlled conditions. The combination of high flow rate and a gradual reduction of the number of animals per culture chamber as they grow makes it possible to keep zooplankton at a constant food concentration throughout their lives. For solutions of the model [Eq. (4.9)] with the functional response given by Eq. (4.11), an incipient limiting food level $C' = 0.17$ (mg C) l^{-1} gave the best fit to a published set of growth curves by Taylor (1985). Figure 4.8 indicates that a constant C', independent of body size, is sufficient to describe food-limited growth in *Daphnia pulex*. The use of a piecewise linear functional response in Eq. (4.11), implying exactly overlapping growth curves for all food levels $C > C'$, finds some support in Taylor (1985), who observed nearly identical growth curves in *Daphnia pulex* at 0.5 and 1.0 (mg C) l^{-1}.

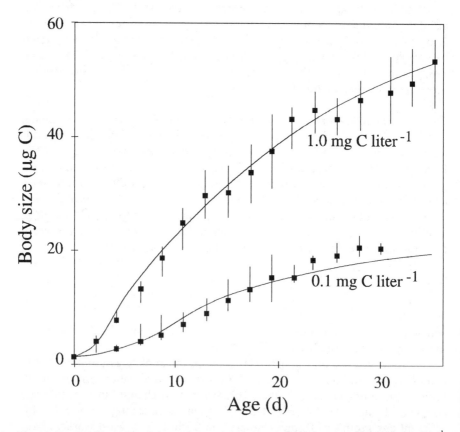

Fig. 4.8. Simulated body growth in *Daphnia pulex* at constant food concentrations of 1.0 (mg C) l^{-1} (*upper curve*) and 0.1 (mg C) l^{-1} (*lower curve*). *Vertical bars* and *squares* are ranges and medians of observations from *Daphnia pulex* grown in flow-trough cultures at the same food levels (Taylor 1985), recalculated to carbon units using Eq. (4.5)

Threshold Food Level for Positive Individual Growth. Lowering the food concentration from the incipient level and downwards will lead to a proportional reduction in the ingestion rate. When the food concentration is so low that assimilation is just adequate to cover the costs of basal metabolism, an animal is said to have reached the threshold food level [C''; (mg C) l^{-1}] for positive individual growth (Lampert 1977). For a set of zooplankton individuals competing for a common food resource, the individual with the lowest C'' will be superior, because it is able to lower the food concentration to a level where the competitors are unable to capture enough food to cover their maintenance costs (DeMott 1989). From the model of Hebert (1982), it is expected that competition among zooplankton will increase with increasing genetic similarity, and that it will be most intense among individuals of the same species sharing the same niche.

Assimilation is just adequate to cover the costs of basal metabolism when $\varepsilon I = r$, or by inserting $C = C'$ in Eq. (4.11) under the necessary assumption that $C'' \leq C'$, when

$$\varepsilon I' \frac{C''}{C'} = r. \qquad (4.12)$$

With C' being assumed independent of body size, Eq. (4.12) implies that C' should increase with body size. Solving Eq. (4.12) with respect to C'' for $C' = 0.17$ (mg C) l^{-1} and $I' = I'_1$ gives a threshold food concentration for positive neonate growth $C''_1 = 0.05$ (mg C) l^{-1} with a $\approx 10\%$ increase from neonate to primiparous adult. Lampert (1977) found a similar size dependence for *D. pulex* feeding on *Scenedesmus*, while the trend was reversed when *Stichococcus* was used as food source. Gliwicz and Lampert (1990) found higher threshold food levels for egg production than for positive juvenile growth in several *Daphnia* species, indicating that juveniles might be competitively superior to their mothers.

Estimates based on short-term radiotracer experiments (Lampert 1977), egg production in natural populations (Lampert and Schober 1980), and growth experiments under constant food conditions (Lampert and Muck 1985) are fairly consistent in indicating a threshold food concentration for individual growth around 0.05 (mg C) l^{-1} for several *Daphnia* species (*D. pulex*, *D. longispina*, and *D. rosea*). On the other hand, species like *D. pulicaria* and *D. hyalina* seem to have threshold levels significantly below 0.05 (mg C) l^{-1} under optimal conditions (Gliwicz and Lampert 1990).

Food-Limited Daphnia Growth in Transfer Culture - a Case Study. Most growth experiments with *Daphnia* under limiting food conditions have been performed with transfer-culture methods; that is, animals are grown individually in small vessels with transfer to fresh food preparations at 1-2-day intervals (Richman 1958; Paloheimo et al. 1982; Lynch et al. 1986; Lynch 1989). A culture volume of 10 ml (as used by Richman 1958) might be expected to be swept clear in a few hours by an adult *Daphnia pulex*, while the same volume may contain sufficient food for several days of growth for a neonate of the same species. Animals grown in semicontinuous culture will therefore not experience the same food concentration throughout their lifetime and the adults will be victim to dramatic "feast and famine" cycles as a result of the feeding regime.

Additional problems will result when the animals are cultured under a light:dark cycle that permits continued growth of the food algae. In the experimental setup of Lynch (1989), this gave a paradoxical extrapolation to positive population growth at zero food concentration. A simple calculation shows that the observed net secondary production exceeded the food ration by nearly an order of magnitude at the lowest food level ("0.0154 µg C ml^{-1}") of Lynch (1989), probably indicating that, in addition to algal growth after transfer, bacterial contamination also contributed to the food supply.

As a case study on *Daphnia pulex* growth in transfer culture, we can take a series of life-table experiments conducted by Frank et al. (1957). They chose to vary the number of animals per vessel, with all vessels receiving the same food addition at 2-day intervals, a design which probably makes it easier to maintain a well-defined food supply among treatments than in experiments where animal density is constant and food concentration is varied. The major weakness with this experimental protocol would nevertheless be the same as for other semicontinuous designs: adults will experience a much more severe food shortage between food replenishments than juveniles.

Frank et al. (1957) reported *Daphnia pulex* body size as biovolume calculated from carapace length and width measurements, instead of the more common units of length or dry weight used in more recent accounts. The average biovolumes of neonates, primiparous females, and the largest females in Frank et al. (1957) were 0.2, 2.3, and 8.7 mm^3. Comparing these biovolumes with the corresponding carbon units in Table 4.1 indicate a roughly linear relationship between biovolume and carbon mass, using a conversion factor of 6 (µg C) mm^{-3}.

In contrast to the flow-through system simulated in Fig. 4.8, changes in food concentration needs to represented explicitly in a model of *Daphnia* growth in transfer culture. If animals are cultured in the dark (as in the experiments of Frank et al. 1957), the mass balance of the food compartment will have no growth term; the body growth model [Eq. (4.9)] therefore needs only to be augmented with the differential equation

$$\dot{C} = -I\,n\,B,   \tag{4.13}$$

describing the reduction in food concentration [C; (mg C) l^{-1}] through consumption. In Eq. (4.13) the specific ingestion rate I is a function of body size and food concentration [Eqs. (4.10), (4.11)], whereas n is the crowding level as individuals ml^{-1}. If body size (B) is in units of µg C, as in Eq. (4.9), the right hand side in Eq. (4.13) will have the correct dimension (µg C) ml^{-1} day^{-1} = (mg C) l^{-1} day^{-1}. Figure 4.9 shows the results of solving Eqs. (4.9) and

(4.13) with transfer and food replenishment being simulated by resetting C every 2 days. Initial food concentration after transfer was calculated from the cell density of *Chlamydomonas moewusi* used by Frank et al. (1957; $9.6 \cdot 10^6$ ml^{-1}), assuming the same carbon content as the related *C. reinhardi* (0.53 pg C cell^{-1}; Lynch et al. 1986).

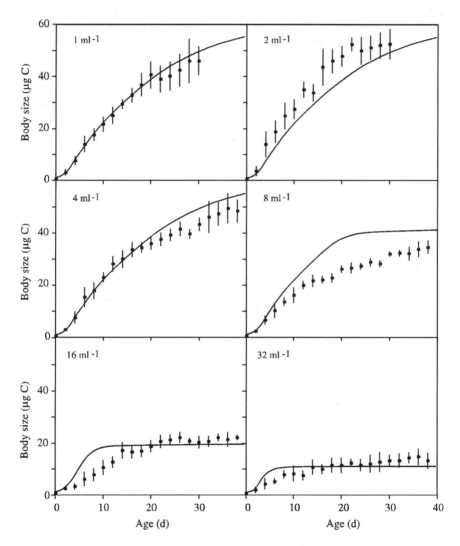

Fig. 4.9. Simulated body size of *Daphnia pulex* in transfer culture at densities of 1, 2, 4, 8, 16, and 32 individuals ml^{-1} (*continuous curves*). *Black squares* and *vertical bars* are means and standard deviations of observations from *Daphnia pulex* raised at the same densities (Frank et al. 1957)

The simulations indicate that growth at the three lowest densities in Fig. 4.9 were indistinguishable from unlimited growth in the experiments of Taylor (1985) and Lynch (1989). Under a semicontinuous feeding regime, all animals will initially grow at the maximum rate and depart from the un-limited growth curve when reaching a size where the energy requirements corresponding to maximum growth rate exceed the ration (Fig. 4.9). Body growth stops when the fraction of the ration energy allocated to somatic growth is used up by respiration and maintenance in the starvation period after the food is exhausted. This does not imply that egg production stops, as one of the main assumptions behind the model is that material allocated to reproduction before the food is exhausted cannot be reallocated to maintenance when the starvation period starts. In contrast, under the constant, suboptimal food conditions of a flow-through system, growth will follow the same pattern as for animals with sufficient food, but the overall growth rate will be reduced (Fig. 4.8).

Food-Limited Individual Growth in Daphnia. In the course of this chapter, a model of individual growth and reproduction in *Daphnia* has been formulated, with an emphasis on the major changes in the fate of assimi-lated energy accompanying maturation. The abundance of experimental data on *Daphnia pulex* makes it possible to use this genus as a model species and to obtain a high degree of consistency between individual para-meter estimates. The conformity between model output and experimental data from several sources indicates that the model captures some of the essential features of *Daphnia* biology, although it cannot be taken as a rigorous validation of the model.

$$\dot{B} = \left(g - (Rg)_+ - h\right)B \tag{4.14}$$

$$R = R'\left(1 - \left(\frac{B'}{B}\right)^2\right)_+ \tag{4.15}$$

$$g = \varepsilon I - r \tag{4.16}$$

$$I = \frac{I_1'}{C'}\,\mathrm{Min}(C, C')\,\frac{B_\infty' - B}{B_\infty' - B_1} \cdot \tag{4.17}$$

The final set of model equations and parameter estimates are displayed in Eqs. (4.14)-(4.17) and Table 4.2 [the function $\mathrm{Min}(x,y)$ is equal to x if $x \le y$ and y otherwise, while the shorthand notation $z = (x)_+$ means that $z = x$ if $x \ge 0$ and $z = 0$ otherwise]. Body growth is dependent on relative expenditures of acquired food carbon into reproduction and molting. Equation (4.14) incorporates the allocation rule that reproductive investment is stopped if the net carbon balance goes negative ($g \le 0$). The cost of molting is a constant

Table 4.2. Model parameters for body growth in *Daphnia pulex*

	Parameter	Value	
B_1	Neonate size	1.1	µg C
B'	Size at first reproductive investment	7.5	µg C
B'_∞	Extrapolated body size where ingestion is zero	132.0	µg C
C'	Incipient limiting food concentration	0.17	(mg C) l^{-1}
ε	Food carbon assimilation efficiency	0.8	–
h	Specific molting loss rate	0.045	day^{-1}
I'_1	Maximum specific ingestion rate in neonates	1.05	day^{-1}
R'	Maximum fraction allocated to reproduction	0.8	–
r	Specific respiration rate	0.25	day^{-1}

fraction of body mass, while the allocation to reproduction increases with body size after the size of the first reproductive investment is reached [Eq. (4.15)]. Both assimilation efficiency and maintenance costs are independent of body size [Eq. (4.16)], while ingestion decreases linearly with body size [Eq. (4.17)].

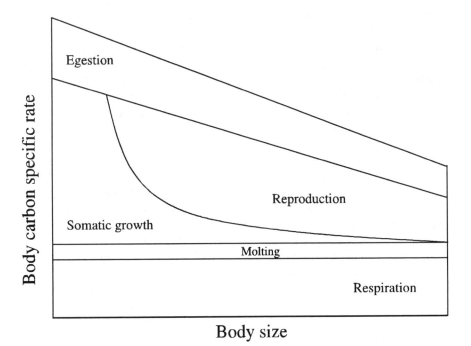

Fig. 4.10. Allocation scheme for ingested food as a function of body size used in the model of growth and reproduction in *Daphnia pulex*

The resulting allocation scheme for ingested food carbon as a function of body size can be summarized by Fig. 4.10. As material lost by molting is taken from the investment in somatic growth, the net effect will be much less in juveniles, which invest all net assimilate in somatic growth, than in adults, where the major fraction of net assimilate is invested in reproduction. The combined effects of decreasing energy input and the increasing relative costs of metabolic and structural maintenance with body size, create the characteristic sigmoidal growth curve in *Daphnia*. The retardation of body growth rate makes body size change very slowly with age in adults, which again leads to the nearly constant net production rate in adult *Daphnia*, as reported by Lynch et al. (1986).

Comparison with Other Daphnia Growth Models. Although the assumption of competitive superiority of large-bodied animals forms the basis of both the competition part of the size-efficiency hypothesis (Brooks and Dodson 1965) and several other more general zooplankton growth models (Hall et al. 1976; Lynch 1977), it seems not to be given strong support by recent investigations (DeMott 1989). Both the present and several other *Daphnia* models (e.g., Kooijman 1986; Gurney et al. 1990; McCauley et al. 1990) predict that juveniles should be competitively superior to adults under food limitation. The model of Paloheimo et al. (1982), with which the present model otherwise shares important structural similarities, predicts the opposite, mainly due to an assumption of increasing assimilation efficiency with body size.

Kooijman (1986) presented a model of *Daphnia* growth and reproduction under food limitation, where the discontinuities associated with molting, maturation, brood production, and starvation in *Daphnia* are deliberately neglected. Even the allocation to reproduction is represented as a smooth function of size under the assumption that juveniles allocate a fraction of their assimilate in preparation of adulthood. As noticed by McCauley et al. (1990), maintaining a continuous dependence of reproduction on assimilation implies that the animals continue to reproduce them selves to death when their overall carbon balance goes negative. On the other hand, constructing a model that is analytical everywhere also offers advantages in the sense that the model is considerably more wellbehaved from a mathematical point of view, allowing analytical treatment to an extent which is most likely unattainable in more realistic models.

Gurney et al. (1990) and McCauley et al. (1990) formulated a *Daphnia* model that gives full attention to the discreteness of *Daphnia* life-history events. By choosing to represent the state of an individual in terms of both carapace length (which changes only at molt) and body weight (which changes continuously between molts), they also obtained the advantage of representing short-term starvation in a simple way as the *weight-for-length* index: an animal is defined as starving if its current body weight is less than would be expected from its carapace length. By defining the onset of maternity in terms of carapace length, they also avoided a rather embarrassing property of other models, including the present one, where maturation is described in terms of body weight; that adults might starve themselves back to immaturity. The weight-for-length index makes the model specially suitable to describe the regular within-instar starvation episodes experienced by animals in transfer cultures, as is well illustrated by the impressive performance of their model against a validation suite based on *Daphnia pulex* culture data. On the other hand, the level of biological detail in the model of Gurney et al. (1990) and McCauley et al. (1990) also has its costs, in both a large number of parameters (18), and numerical complications arising from representing clutch production and molting as discrete events. The latter problem seems to have hindered McCauley and coworkers from extending their model from single individuals to whole populations; in their only attempt on *Daphnia* population modeling so far (Nisbet et al. 1989), they resorted to a delay-differential equation formulation sacrificing much of the biological insight gained in their previous models.

The present model represents an intermediate level of complexity between extremes described above, and as such, it contains tradeoffs which affect both analytical tractability and predictive power compared to the models of Kooijman (1986), and Gurney et al. (1990) and McCauley et al. (1990). Such compromises, based on sound biological judgement, seem unavoidable in constructing a practical model of natural zooplankton populations within the constraints given by realism, experimental evidence, and structural and computational complexity.

4.5 The Economy of Essential Elements

Although some workers have demonstrated direct uptake of inorganic ions like phosphate in *Daphnia* (e.g., Parker and Olson 1966), this flux will probably only be of any importance at concentrations several orders of magnitude above what is common in natural waters (Peters 1987). It is therefore likely that aquatic herbivores are dependent solely on their food to obtain the amounts of biomolecules and essential elements that are necessary to support life.

The Importance of Elemental Composition in Determining Food Quality.
Most of the work on nutrition and growth limitation in *Daphnia* and other zooplankton has focused on the effects of food quantity, while food quality usually has been associated with differences among functional or taxonomic groups (e.g., Arnold 1971; Geller 1975; Lampert 1977; Pace et al. 1983; Infante and Litt 1985). More recently, a growing body of evidence shows that other aspects, like chemical and biochemical composition, can also be of vital importance to the nutritional quality of food.

Checkley (1980) found that the gross growth efficiency (the fraction of ingested food incorporated into gonadal or somatic tissue) of the marine copepod *Paracalanus* was related to the N:C ratio of the food. Apparently, the "excess" carbon in algal food with low N:C ratio was disposed of as increased feces production, indicating that major changes in food carbon utilization can result from changes in the elemental composition of the food.

Lehman and Naumoski (1985) reported large differences in phosphorus assimilation and excretion among *Daphnia pulex* raised on P-limited and P-sufficient algal food, suggesting that the phosphorus utilization of the animals is also affected by the P content of their food. Olsen et al. (1986b) inferred from phosphorus mass-balance studies on *Daphnia* that there must be a critical P:C ratio below which the animals are unable to sustain their maximal growth rate, and showed that estimates of this critical food phosphorus content were within the P:C ratios observed in mildly nutrient-limited algae. The idea of nutrient-limited zooplankton growth is further supported by Sterner et al. (1993), who found dramatic reductions in growth, reproduction, and survival in *Daphnia pulex* reared on P-limited *Scenedesmus acutus* compared to animals given nutrient-sufficient cells of the same food species.

Andersen and Hessen (1991) concluded from an investigation of C, N, and P content in natural populations of several species of zooplankton that interspecific variations in elemental composition were generally higher than intraspecific variations, and that this pattern was preserved under experimental starvation and food enrichment, suggesting that zooplankton are able to maintain their elemental ratios under changing food composition. This homeostatic view of zooplankton mineral nutrient economy is also supported by Sterner (1990), who showed that, among several possible alternatives, a model based on the assumption of constant elemental ratios in the grazers gave the best fit to empirical data from Le Borgne (1982) on N and P release by marine zooplankton.

Growth Limitation by Food P Content. The maintenance of constant elemental composition puts strong constraints on the fraction of food C that can be utilized for production of new biomass when the food is P-deficient relative to the requirements of the animal. In Appendix A3 it is shown under quite general conditions that the constraints from balanced growth will be satisfied when the P loss rate from catabolic processes is described by a family of power functions [Eq. (A3.11)], enabling different degrees of catabolite reutilization to be represented by varying the exponent n. Different strategies for nutrient conservation give relationships between food P content and food C assimilation efficiency (Fig. A3.3) which are confined between the two extremes corresponding to constant excretion rate ($n = 0$), and to zero excretion rate when maximal C assimilation is impossible ($n = \infty$). By insertion into the carbon budget equation (4.2) of the carbon-specific assimilation efficiencies corresponding to the three candidate strategies identified in Appendix A3 [Eqs. (A3.14)-(A3.16], the following growth models can be proposed for animals maintaining balanced growth on nutrient-deficient food under different degrees of nutrient conservation efficiency:

$$g_0 = g^* \left(1 - \left(1 - (Q/\theta) \right)_+ / K_2^* \right) \tag{4.18}$$

$$g_1 = g^* \left(1 - \left(1 - (Q/\theta) \right)_+ \right) \tag{4.19}$$

$$g_\infty = g^* \left(1 - \left(1 - (Q/\theta)/K_2^* \right)_+ \right) \tag{4.20}$$

where Q and θ are the P contents of food particles and animals [$(\mu g\ P)(mg\ C)^{-1}$], $g^* = \varepsilon^* I' - r$ is the maximum growth rate at saturating levels of optimally composed food, and $K_2^* = g^*/(\varepsilon^* I')$ is the corresponding maximum net growth efficiency.

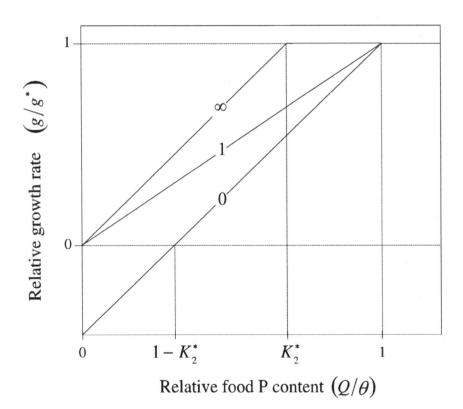

Fig. 4.11. Hypothesized relationships between carbon-specific growth rate and food composition for three different nutrient utilization strategies. *Curves* labeled 0, 1, and ∞ are given by Eqs. (4.18)-(4.20)

Figure 4.11 shows how the hypothesized growth functions are related to each other. All the models given by Eqs. (4.18)-(4.20) will be identical to Eq. (4.2) for $Q \geq \theta$. The least efficient utilization strategy ($n = 0$), exhibits a threshold for positive growth at $Q/\theta = 1 - K^*{}_2$, while the other two ($n = 1, \infty$) enables non-negative growth for all $Q \geq 0$. The most efficient utilization strategy ($n = \infty$) will enable maximum growth rate to be maintained for $Q/\theta \geq K^*{}_2$, while the other two ($n = 0, 1$) will have reduced growth rate for $Q < \theta$.

The Phosphorus Utilization Strategy in Daphnia: No successful attempts have so far been made to determine the growth rate of freshwater zooplankton on food with a well-defined and reproducible P content. Any inference about which of the curves in Fig. 4.11 (if any of them) might describe P-limited zooplankton growth must be made through an indirect approach.

In Chapter 2, phosphorus recycling by zooplankton was described by the carbon-specific release rate ρ [(μg P) (mg C)$^{-1}$ day^{-1}], which is determined by the difference between P ingestion and P utilization (Olsen and Østgaard 1985):

$$\rho = I\,Q - g\,\theta. \tag{4.21}$$

When the P-limited growth models [Eqs. (4.18)-(4.20)] are substituted into Eq. (4.20), we obtain the following set of expressions for the P release rate:

$$\rho_0 = I\theta \left((Q/\theta) - \varepsilon^* \left(K_2^* - (1 - (Q/\theta))_+ \right) \right), \tag{4.22}$$

$$\rho_1 = I\theta \left((Q/\theta) - \varepsilon^* K_2^* \left(1 - (1 - (Q/\theta))_+ \right) \right), \tag{4.23}$$

$$\rho_\infty = I\theta \left((Q/\theta) - \varepsilon^* \left(K_2^* - (K_2^* - (Q/\theta))_+ \right) \right), \tag{4.24}$$

All the expressions (4.22)-(4.24) describe piecewise linear functions with positive slopes. For $Q < \theta$, the three candidate models can be ranged as $\rho_0 > \rho_1 > \rho_\infty$, in accordance with the different degrees of nutrient conservation represented by the three strategies. For $Q \geq \theta$, all three functions will be identical and equal to $IQ - g^*\theta$. It should also be noticed that all the equations (4.22)-(4.24) can be reexpressed in two dimensionless variables, $\rho/I\theta$ and Q/θ, and two parameters, ε and K_2'.

The measurement of zooplankton nutrient recycling is still connected with large methodological problems (Lehman 1984; Sterner 1989), and very few workers have so far studied nutrient release in terms of zooplankton nutrient utilization, as introduced by Olsen and Østgaard (1985). With the notable exception of Olsen and coworkers (Olsen and Østgaard 1985; Olsen et al. 1986b), very few studies have attempted to measure phosphorus release and carbon ingestion simultaneously, which is necessary to compare the relative merits of Eqs. (4.22)-(4.24) in describing zooplankton P release.

If we consider the experiments of Olsen et al., where a defined algal food source is used (*Scenedesmus* or *Rhodomonas*), we can with some certainty assume a maximal carbon assimilation efficiency $\varepsilon^* = 0.8$, as discussed in Section 4.3. Under this assumption, the growth model [Eqs. (4.16), (4.17)] gives a maximal net growth efficiency $K^*_2 = 0.7$ for *Daphnia* with a body size of 23 µg C (the mean of the *D. pulex* specimens used by Olsen and coworkers). From these dimensionless parameters we can predict the relationship between the dimensionless recycling index $\rho/I\theta$ and the relative food P content Q/θ for the three candidate models, Eqs. (4.22)-(4.24) (Fig. 4.12).

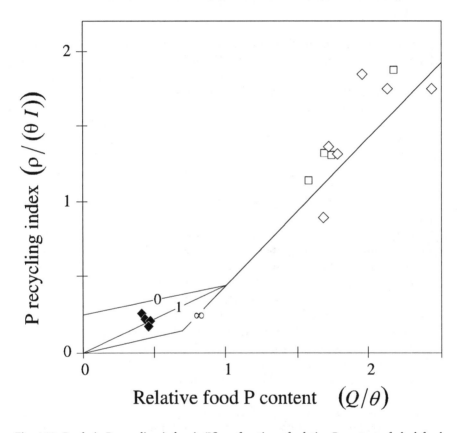

Fig. 4.12. *Daphnia* P recycling index ($\rho/I\theta$) as function of relative P content of algal food (Q/θ). *Black lines* marked 0, 1, and ∞ are the predicted relationships from Eqs. (4.22)-(4.24); *filled diamonds* D. pulex grazing on *Scenedesmus*; *open squares* D. longispina grazing on *Rhodomonas* (both from Olsen and Østgaard 1985); *open diamonds* D. pulex grazing on *Rhodomonas* (Olsen et al. 1986b)

Olsen et al. (1986b, Fig. 6) reported P released per C ingested (ρ/I) as function of the P:C ratio of the food particles (Q). If we rescale these observations by the typical P content in *Daphnia* [$\theta = 30$ (μg P) (mg C)$^{-1}$; Olsen et al. 1986b; Andersen and Hessen 1991], the P recycling index $\rho/I\theta$ and the relative P content of the food algae Q/θ can be calculated. Although the available data are limited (specially for an interval around $Q/\theta = 1$), Fig. 4.12 indicates that the intermediate P utilization strategy ($n = 1$) fits the observations better than the other two ($n = 0$ or ∞). When it is kept in mind that the parameters $\overset{*}{\varepsilon}$ and K^{*}_{2} are estimated independently of the data, the correspondence between observations and predictions must be considered satisfactory.

Growth Limitation by Food Composition and/or Abundance. So far, we have implicitly assumed that food abundance is sufficient to support a positive carbon balance in the animals. Andersen and Hessen (1991) showed that cladoceran zooplankton were able to maintain constant P:C ratio even when starving, indicating that body P is lost in proportion to respiratory C losses when the overall carbon balance is negative. If the food carbon concentration is below the threshold level for positive individual growth ($C < C''$), so that the animals have a net loss of body C, it is reasonable to assume that growth (loss) is independent of food composition. In other words, that growth is proportional to food P content only when $C < C''$ and $Q < \theta$; otherwise growth is determined by the carbon balance alone.

In the discussion above, we have considered only the maintenance costs of respiration, and neglected the body material lost by molting. This approximation will be valid if we assume that the molt has the same P:C ratio as the rest of the body. While Andersen and Hessen (1991) found the exoskeleton residues after persulfate digestion to contain no phosphorus, this does not necessarily mean that the intact exoskeleton is P free. On the contrary, Yan et al. (1989) indicate that the *Daphnia* carapace has a high P content, while Scavia and McFarland (1982) found conspicuous peaks in the P release rate from *Daphnia magna* at the time of molting, indicating that the P contained in the old carapace cannot be completely reclaimed for anabolic purposes.

If we assume that molting represents a proportional loss of body P and C, the maintenance of a constant P:C ratio should have no effect on the allocation between growth, reproduction, and molting. Food composition then will only affect the total amount of material available for producing new biomass as body tissues, exoskeleton, or eggs. This means that Eqs. (4.14), (4.15) will be unmodified by the inclusion of food composition effects in the individual growth model of Section 4.4, as will the submodel describing feeding as function of body size and food concentration [Eq. (4.17)]. In order to describe the carbon balance under suboptimal food composition, Eq. (4.16) needs to be replaced by the following pair of equations:

$$g = g' - (g')_+ \left(1 - (Q/\theta)\right)_+$$ (4.25)

$$g' = \varepsilon^* I - r.$$ (4.26)

Equation (4.26) gives the net assimilation rate g' that would result if food composition were optimal ($\varepsilon = \varepsilon^*$), while Eq. (4.25) contains the reduction in realized growth when food is P deficient. Eq. (4.25) is formulated so that growth will be unaffected by food P content if the total carbon intake is insufficient to maintain positive growth.

If P [(μg P) l^{-1}] is the concentration of P contained in food particles and C [(mg C) l^{-1}] is the food C concentration, food composition will be suboptimal whenever $P < \theta C$, while food abundance will be limiting when $C < C'$. Figure 4.13 shows isolines of constant growth rate as function of food P and C concentration. The surface will be flat, with $g = g^*$ in the region where growth is neither C- or P-limited ($P > \theta C$ and $C \geq C'$), while the isolines will be parallel to the P axis, and thus independent of food composition, when the food C level is insufficient to support positive net growth ($C < C''$)

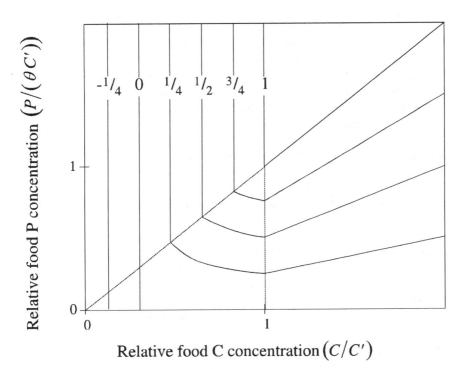

Fig. 4.13. Isolines of constant relative growth rate (g/g^*; *labeled curves*) as function of relative food C and P concentration [C/C' and $P/(\theta C')$]. *Broken lines* indicate the discontinuities at the incipient limiting food concentration ($C = C'$), and at the optimal food P content ($P/C = \theta$)

4.6 Individual Mortality and Population Losses

In natural zooplankton populations, mortality losses are composed of at least two component parts; mortality due to vertebrate and invertebrate predation, and nonpredatory mortality due to starvation and senescence. The impact of predatory mortality on zooplankton populations has been shown to be strongly dependent on the type of predator involved; carnivorous crustaceans and insects exhibit a selective mortality on small species and/or juvenile stages, while planktivorous stages of fish and amphibians have a strong selectivity for large species and/or adults (Zaret 1980; Kerfoot and Sih 1987; Gliwicz and Pijanowska 1989).

Since the pioneering work of Holling (1966), much progress has been made in identifying the component processes of predation, where a successful capture of a prey is described by a Markov chain of probabilities for encounter, attack, and ingestion (e.g. Williamson and Gilbert 1980). The encounter probabilities have been accurately modeled as functions of prey visibility and the swimming speeds of prey and predator by Gerritsen and Strickler (1977). Compared to the detailed studies carried out by numerous workers on predatory mortality in *Daphnia* and other zooplankters, the processes involved in mortality from starvation and senescence have received less attention and are generally much less well understood. As noticed by Lynch (1983, 1989), reliable size-specific mortality rates in natural *Daphnia* populations are almost completely lacking, so that the main source of information on nonpredatory mortality rates is from life-table experiments under laboratory conditions.

Daphnia Survival Under Food-Sufficient Conditions. While mortality most certainly is a binary process on the individual level (an individual is either alive or dead), it is usually observed on the population or cohort level as a survival curve describing the fraction remaining alive as the initial cohort is aging. In the typical life-table experiment, a cohort of newborn is raised under identical conditions while the state of each individual (dead or alive) is being monitored. In Appendix A4 it is shown that the empirical survival curve of a finite cohort is an unbiased and consistent estimator for the true survival function of the population. The survival function relates the accumulated mortality loss on the population level to the instantaneous mortality rate (or force of mortality) on the individual level.

The problems associated with maintaining animals at a constant food level throughout their lives, as discussed in Section 4.4, suggest an approach to mortality modeling similar to the model of growth and reproduction developed earlier in this chapter: to start with a model of the mortality process in populations growing at nonlimiting food levels and then try to expand this model to situations where survival is modulated by food limitation and starvation. The zooplankton literature contains a rich source of survival experiments from several species of *Daphnia*. If we continue to use *Daphnia pulex* as our model species, we can, in addition to life-table experiments performed in ecological contexts, like Lynch (1989), find a large base of relevant *Daphnia* survival data within the ecotoxicology literature, either as control treatments or treatments where no toxic effect was detected. Combining the survival curves from the experiments by Lynch (1989), where population growth was found to be independent of food level [>1 (mg C) l^{-1}], with survival curves from the treatments in Winner and Farrell (1976) and Daniels and Allan (1981) where no toxic effects were found, gives a set of 12 survival curves for *Daphnia pulex* under sufficient food supply.

Figure 4.14 suggests that most *Daphnia pulex* individuals die either as early juveniles (<5 days old) or as old adults (>25 days old), with the survival curve at intermediate ages being nearly flat. Such a survival pattern seems to be common in both human demography (Lawless 1982) and reliability engineering (Hahn and Meeker 1982), and has been inter-

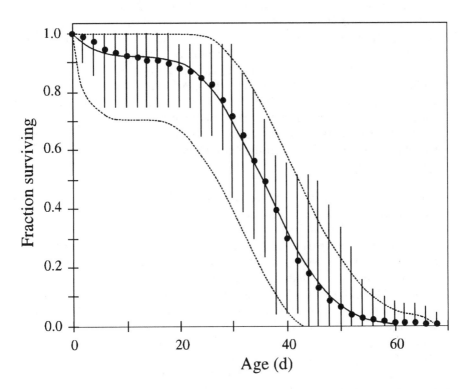

Fig. 4.14. Survival in *Daphnia pulex. Vertical bars* are ranges from a set of 12 *D. pulex* survival curves reported by Winner and Farrell (1976), Daniels and Allan (1981) and Lynch (1989). *Solid line* is the fitted survival function from the bathtub model, given by Eq. (4.26); *broken lines* delimit the 95% confidence region for this survival function, as given Eqs. (A4.13) and (A4.14), for an initial cohort size of 20 individuals

preted as a purging of weak individuals, leaving a reduced population with higher survival probability. This kind of survivorship is usually well described by a convex or "bathtub"-shaped (Lawless 1982) force of mortality $\eta(x)$ (see Appendix A4 for details). The simplest possible force of mortality with this convexity property can be written as

$$\eta = \eta_0 \left(1 - x/x'\right)^2,$$
(4.27)

with η_0 being the instantaneous mortality rate (day^{-1}) at age $x = 0$, and x' the age at which $\eta = 0$. The corresponding survival function is given by substituting Eq. (4.27) into Eq. (A4.5) and solving the integral in the exponent:

$$l(x) = \exp\left(-\tfrac{1}{3}\eta_0\, x'\left(1 - (1 - x/x')^3\right)\right)$$
(4.28)

When Eq. (4.28) is fitted to the survival curves shown in Fig. 4.14, the parameters of Eq. (4.27) are estimated as $\eta_0 = 0.0226 \pm 0.0060$ day^{-1} and $x' = 12.6 \pm 1.2$ days. In other words, approximately 2% of the neonates die per day, while mortality is practically zero at an age around 12 days; 50% of the population die before reaching an age of 35.9 days (the median survival time).

The deviations from the average survival function within each cohort did not seem to vary systematically among experimenters or treatment levels. Fig. 4.14 shows that much of the variability among individual survival curves is within the expected range for the small initial cohort sizes (mostly 20 individuals) used in these experiments. Although more than 5% of the survival scores in Fig. 4.14 fall outside the binomial 95% confidence limits, the presence of additional within-treatment variance as shown in Appendix A4 indicates that these confidence limits are not sufficiently robust to reject the hypothesis that all the survival curves in Fig. 4.14 are members of the same population.

Daphnia Survival in Food-Limited Transfer Culture. In experimental setups with continuous food supply, such as Taylor (1985), the total number of animals that can be handled in a single run is much lower than what is possible in transfer culture designs. This is probably the reason why apparently all published survival curves for *Daphnia* have been measured in transfer culture experiments (e.g., Ingle et al. 1937; Frank et al. 1957; Goulden et al. 1982; Porter et al. 1983; Lynch 1989). This means that all data on *Daphnia* survival under limiting food conditions might suffer from the same limitations as discussed in Section 4.4 (cf. Tillmann and Lampert 1984; DeMott 1989).

Survival curves for *Daphnia* populations raised at low food levels generally show increased juvenile mortality in addition to reduced longevity (Frank et al. 1957; Goulden et al. 1982; Lynch and Ennis 1983; Porter et al. 1983; Lynch 1989). Juvenile mortality has been proposed to result from a lower resistance to withstand starvation than in adults, but this cannot be the full explanation as juveniles in transfer-culture experiments receive a much higher per capita food supply than adults (cf. Section 4.4).

As mentioned in Section 4.1, two major components of *Daphnia* eggs, oocytes and yolk precursors, are kept separate until the later stages of oocyte maturation. Tessier et al. (1983) raised *Daphnia* to adulthood at high food levels, and began starving different groups of animals from 0 to 60 h after entering the third adult instar. Their experiment showed a linear relationship between clutch size and the fraction of the instar spent at high food levels, indicating that allocation to oocyte material is irreversible. Goulden et al. (1984) found that neonates from *Daphnia magna* raised in transfer culture under food limitation had reduced levels of triacylglycerol (TAG) lipids, indicating that, contrary to oocytes, the material accumulated as energy reserve for the eggs can be reclaimed by the mother under conditions of acute starvation.

Taken together, these pieces of information could indicate that oocyte material and yolk precursors are allocated in proportion during the initial part of the transfer cycle when food is ample, while some of the yolk material is used for maternal maintenance during the starvation phase of the cycle. The reduced TAG lipid content of neonates raised under low food conditions would then be mainly an artifact created by the feeding regime in transfer cultures through the implied decoupling of the allocation of oocyte material and energy reserves. If a female is able to reclaim material allocated to yolk precursors in the starvation part of the transfer cycle, there will be less yolk material available to the oocytes produced in the same brood cycle. If yolk and oocytes are combined serially as the oocytes pass through the oviducts (Zaffagnini 1987), the insufficient total amount of yolk material might be divided so that the first eggs to enter the oviducts receive their full share, while the rest have none.

Lynch and Ennis (1983) found that *Daphnia pulex* neonates, from mothers raised at low and high food levels, showed significant differences in both juvenile mortality and adult longevity when raised at the same food level. The observed maternal effect on juvenile survival is consistent with the proposed decoupling of oocyte and yolk precursor accumulation in food-limited transfer cultures, while the effect on longevity is probably due to factors other than neonatal energy reserves. Lynch and Ennis (1983) also found increased juvenile growth rates in offspring of mothers raised at high food levels, indicating that early juvenile growth is partly dependent on maternally supplied resources, as also suggested by Goulden et al. (1984). If maternally supplied resources are of such vital importance to juvenile survival as indicated by Tessier et al. 1983, it is more likely that juvenile mortality under food limitation is an effect of maternal starvation, and therefore to a large extent a result of the feeding regime in transfer cultures. The fact that most workers have been careful to precondition the animals, with experiments being started after several generations at the same food level, would tend to enforce such a maternal effect on juvenile mortality.

Under this assumption, the clutch of a starving female should be composed of eggs with both normal and with suboptimal yolk content. A fraction of the eggs would then be expected to develop into neonates with normal survival probability, while the other fraction would suffer a higher mortality rate from the lack of some maternally supplied, essential material. If the fraction of aberrant eggs is denoted by α, the composite survival function $l(x)$ of these two subpopulations can be written as

$$\bar{l}(x) = \alpha l'(x) + (1-\alpha)l(x),$$ (4.29)

with $l'(x)$ and $l(x)$ being the survival functions of individuals developing from aberrant and normal eggs, respectively. If we assume that individuals developing from aberrant eggs have a constant mortality rate η' (day^{-1}), while the rest of the cohort have the survival function [Eq. (4.28)], then Eq. (4.29) can be written as

$$\bar{l}(x) = \alpha \exp(-\eta' x) + (1-\alpha)\exp\left(-\tfrac{1}{3}\eta_0 \, x' \left(1 - (1 - x/x')^3\right)\right).$$ (4.30)

Both $l'(x)$ and $l(x)$ are considered generic properties of each egg type, so that the composite survival function from each feeding regime in a set of experiments should be determined by a single parameter α, while the mortality rate η' should be common parameter for all treatments.

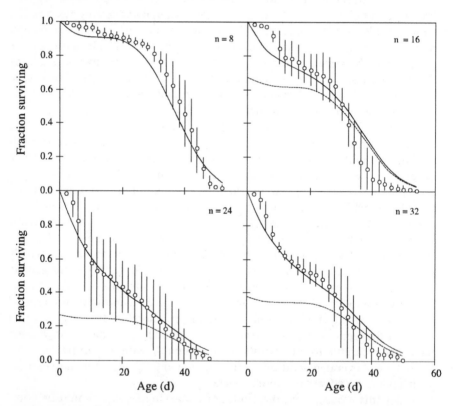

Fig. 4.15. *Daphnia pulex* survival in food-limited transfer culture. *Open circles* and *vertical bars* are means and standard deviations of survival curves from Frank et al. (1957) at densities of 8, 16, 24, and 32 individuals ml^{-1}. *Solid lines* are fitted composite survival functions given by Eq. (4.30); *broken lines* are survival functions of the normal subpopulations

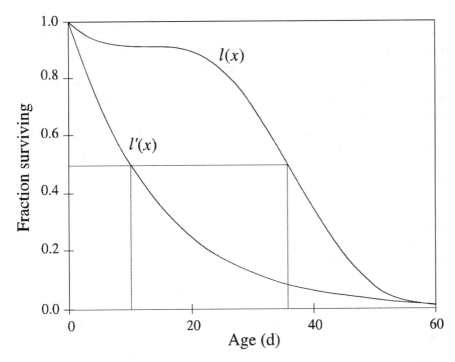

Fig. 4.16. Survival curves for the two subpopulations in the composite model [Eq. (4.30)], as estimated from the data of Frank et al. (1957) (Fig. 4.15). *Broken lines* denote the median survival times of the two distributions

Figure 4.15 shows composite survival curves given by Eq. (4.30), fitted to those of the experiments of Frank et al. (1957) where Fig. 4.9 indicated food-limited growth (densities ≥ 8 ind. ml^{-1}). Estimating a constant mortality rate for aberrant individuals from all four crowding levels shown in Fig. 4.15 gave $\eta' = 0.07 \pm 0.01$ day^{-1}. In Fig. 4.16 the survival functions for the two subpopulations in the composite model [Eq. (4.30)] are displayed; the median survival time for deviating individuals is only 27.5% (9.9 days) of normal individuals (35.9 days). For the lowest density (8 ind. ml^{-1}), the survival curve is similar to the one for food-sufficient animals (Fig. 4.14), indicating that this level of food limitation had no effect on egg and neonate quality. For the higher densities, the fraction of aberrant individuals (α) increases from 0.34 ± 0.05 at 16 ind. ml^{-1}, to 0.77 ± 0.10 and 0.66 ± 0.09 for densities of 24 and 32 ind. ml^{-1} (the last two being not significantly different).

The fitted curves in Fig. 4.15 illustrate how survival curves from food-limited transfer cultures can be decomposed into two subpopulations with different survival properties, where one of the subpopulations apparently has a mortality schedule indistinguishable from food-sufficient populations. Although the fraction of eggs developing into individuals with normal survival properties decreases with decreasing food level, no attempts have been made to develop any more detailed model of the relationship between reallocation of material in the starvation part of the transfer cycle and the quality of eggs laid in the next clutch.

In the course of this section, we have presented evidence that *Daphnia* survival in food-limited transfer culture might be dominated by maternal effects to an extent where the death of an individual bears little relationship to the food level the individual itself has experienced throughout its lifetime. The maternal effects seem to be caused by the decoupling of oocyte and yolk precursor accumulation in the dramatic "feast and famine" cycles experienced by adults at low food treatments in transfer culture, and, as such, mostly artifacts caused by the time scales enforced by the experimental design. Such decoupling of oocyte and yolk precursor accumulation might not be common when food levels change on the time scales experienced by natural populations, as indicated by the results of Tessier and Goulden (1982), who found lipid levels and clutch sizes in natural populations of *Daphnia* and *Holopedium* to be closely correlated.

This interpretation of survival curves from transfer culture experiments calls for extreme caution when extrapolating to natural populations, and implies that we have virtually no reliable information on how survival might be affected by long-term food limitation in natural populations. A parsimonious model might then be that *Daphnia* mortality in natural populations is a food level-independent process that is described by a convex age-dependent function like Eq. (4.27), at least as long as food is sufficient to support positive net growth.

4.7 Population Survival and Proliferation: Demography

The fate of animal populations is determined by the survival and fecundity of its individual members; a female contributes nothing to population growth if she dies before reaching maternity. The egg production rate of a female in the *Daphnia* growth model [Eqs. (4.14)-(4.17)] is a function of animal size and food level. If we assume that an experimental arrangement can be made such that food level can be held at a constant level while a population is expanding (even though this might be complicated in practice), there will be a fixed relationship between age and body size, implying that fecundity can be regarded as a function of female age alone. If both the survival and fecundity schedules at a given food level are age-dependent functions, standard demographic theory (e.g., Pollard 1973; Nisbet and Gurney 1982; or Frauenthal 1986) proves that the population will eventually converge to a stable age distribution with all age classes increasing or decreasing exponentially at the same rate λ (day^{-1}), called the intrinsic rate of increase (see Appendix A5).

Fecundity and Food Limitation. At a given food level C [(mg C) l^{-1}], the solution of the *Daphnia* growth model [Eqs. (4.14)-(4.17)] will give as output body size, net assimilation rate, and fraction of net assimilate allocated to reproduction [$B(x)$, $g(x)$, and $R(x)$,] as functions of age (x). The maternity function, $m(x)$, is defined as the number of offspring produced by a female of a given age in a unit of time, or the net allocation rate to egg production (R g B) divided by the mass of a single egg (B_0):

$$m(x) = \frac{R(x)\,g(x)\,B(x)}{B_0}. \tag{4.31}$$

Figure 4.17 shows how the maternity function [calculated by Eq. (4.31)] is influenced by food limitation in the *Daphnia* model [Eqs. (4.14)-(4.17)]. Decreasing the food level reduces both the juvenile growth rate, so that the age of first reproductive investment is delayed, and the adult reproductive output, so that fewer eggs are produced. Above the incipient limiting food concentration, C', reproductive output will be independent of food level. When the food level is so low that growth is halted by the costs of molting and maintenance before reaching the size of first reproductive investment,

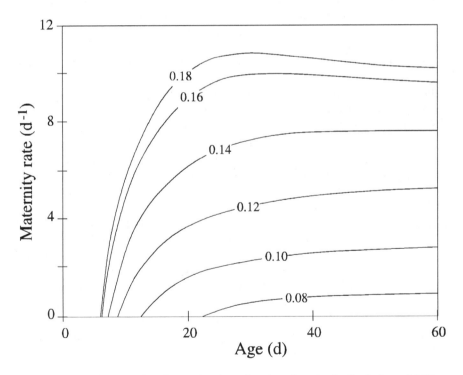

Fig. 4.17. Maternity rate, given by Eq. (4.31), as function of age in the *Daphnia* model [Eqs. (4.14)-(4.17)] for different food levels (*curve labels* (mg C) l^{-1})

the maternity function will be identically zero everywhere. This threshold food level for reproduction, C''_R, can be found by setting $\dot{B} = 0$ and $B = B'$ in Eqs. (4.14)-(4.17) and solving for C:

$$C''_R = \frac{r+h}{\varepsilon \, I'_1} \frac{B'_\infty - B_1}{B'_\infty - B'} \, C' \, , \tag{4.32}$$

or $C''_R = 0.063$ (mg C) l^1 with the parameter values in Table 4.2.

Asymptotic Population Properties - the Intrinsic Rate of Increase. In Appendix A5 it is shown that, from a given pair of maternity and survival functions [$m(x)$ and $l(x)$], the corresponding asymptotic net population growth rate (λ), can be found by solving the characteristic equation.

$$\int_0^\infty m(x)\, l(x)\, e^{-\lambda x} dx = 1 \cdot \tag{4.33}$$

Repeating this procedure for maternity functions generated by a range of food concentrations gives the relationship between net population growth rate and food level (Fig. 4.18). Due to the piecewise linear functional response [Eq. (4.11)], all population statistics will be constant for food levels above the incipient limiting concentration ($C \geq C'$), giving a maximal intrinsic rate of increase $\lambda' = 0.381$ day^1. The maximal intrinsic rate of increase predicted by the model compares well with an average $\lambda' = 0.37 \pm 0.06$ day^1 estimated from a number of life-table experiments on different *Daphnia* species, as summarized in Table A10.7.

Population growth rate falls off quite linearly from the incipient limiting level, and then drops abruptly when food concentration approaches the singularity at the threshold level for reproduction (C''_R). At a food concentration somewhat above C''_R, births and deaths are exactly balanced in the stable age distribution, so that net population growth will be zero ($\lambda = 0$). This threshold food level for positive net population growth (C''_λ) can be found more precisely by setting $\lambda = 0$ in the characteristic equation (A5.5), and locating the value of C corresponding to a net maternity rate of unity [Eq. (A5.6)]. Using the same root-finding procedure as above, the threshold food level for net population growth was found to be $C''_\lambda = 0.0725$ (mg C) l^1.

For two or more zooplankton populations exploiting a common food resource, the population with the lowest threshold food level will be competitively superior, because it will eventually depress the food resource to a level where the other populations are unable to maintain positive net population growth. As conjectured by Lampert (1977), the food level (C''_λ) needed for positive population growth will be higher (16% in this case), than is necessary to reproduce (C''_R).

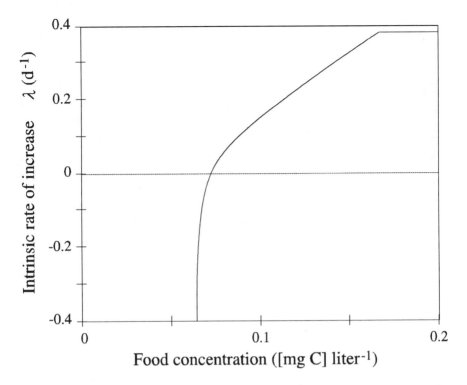

Fig. 4.18. Intrinsic rate of increase as function of food concentration in the *Daphnia* model [Eqs. (4.14)-(4.17)], from solving the characteristic equation [Eq. (4.33)] for the maternity and survival functions given by Eqs. (4.31) and (4.28)

Asymptotic Population Properties - Population-Specific Birth and Death Rates. Lotka (1907) showed that the asymptotic net growth rate of the total population can be interpreted as $\lambda = \beta - \delta$, where β and δ are the specific birth and death rates of the total population after it has reached the stable age distribution (both in units of day[-1]). As shown in Appendix A5, the specific birth rate (β) orresponding to a given net population growth rate (λ) is determined by the survival function $l(x)$ alone:

$$\beta = \left[\int_0^\infty l(x)e^{-\lambda x}dx \right]^{-1}. \tag{4.34}$$

When β has been computed for a given λ, the specific death rate (δ) of this stable age distribution can be calculated simply as the difference $\delta = \beta - \lambda$. Figure 4.19 shows that the specific birth rate β is linearly related to the intrinsic rate of increase for $\lambda > 0$, and decreases asymptotically to zero as λ becomes progressively more negative. At saturating food levels, the specific death rate $\delta' = 0.0196$ day[-1], so that approximately 2% of the population

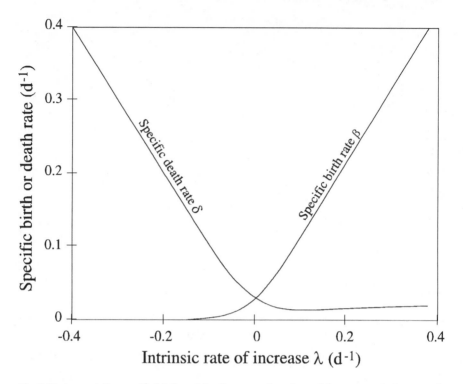

Fig. 4.19. Asymptotic, specific birth and death rates as functions of the net population growth rate (intrinsic rate of increase). Birth rate (β, day^{-1}) calculated from Eq. (4.34) and the survival function [Eq. (4.28)], and death rate (δ, day^{-1}) as $\delta = \beta - \lambda$

members die per day. The specific death rate is nearly constant as long as $\lambda > 0$, while it is linearly related to the net population growth rate in decreasing populations.

Asymptotic Population Properties - the Stable Age Distribution. Demographic theory tells us that when a population has converged to the stable age structure, the stable age frequency distribution can be interpreted as the survival function $l(x)$ depreciated by the intrinsic rate of increase (λ) (see Appendix A5). The cumulative age frequency distribution describing the fraction of population members aged $\leq x$ [corresponding to the distribution in Eq. (A5.10)] can be written as

$$F_n(x) = \beta \int_0^x l(\xi) e^{-\lambda \xi} d\xi, \qquad (4.35)$$

where β is the asymptotic birth rate [Eq. (4.34)]. Figure 4.20 shows cumulative age distributions resulting from the *Daphnia* survival model [Eq. (4.28)] for different net population growth rates. It is seen that in fast-growing populations <10% of the members will be adults, with the median

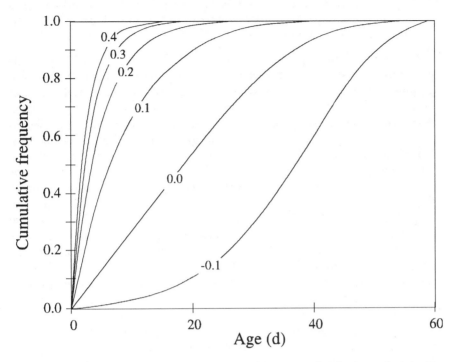

Fig. 4.20. Cumulative plots of the asymptotic age frequency distributions, given by Eq. (A5.10), for the survival function [Eq. (4.28)] and different net population growth rates (intrinsic rates of increase: *curve labels* as day^{-1})

age of the population increasing as the population growth rate decreases. Neonates will be the mode of the age distribution as long as net population growth is positive; for decreasing populations the mode is relocated to older age classes.

Parameter Sensitivities of Asymptotic Population Properties. From the discussion above, we can identify three single-valued asymptotic population properties that can be closely related to the performance of a zooplankton population in certain situations:

1. The maximum intrinsic rate of increase (λ): this generally accepted measure of evolutionary fitness is also related to the innate capacity of a zooplankton population to track the dynamics of their prey, and thus to exploit bursts of food availability.

2. The asymptotic death rate in a food-sufficient population (δ): on a sufficiently long time scale, zooplankton populations will be net sinks of nutrients; zooplankton mortality will therefore be closely linked to the nutrient economy of the whole plankton community.

3. The threshold food level for positive net population growth (C''_λ): a measure of competitive ability toward other zooplankton populations exploiting the same food resource.

These population properties are derived from coupled equations involving a total of 11 parameters describing both energetic and demographic aspects of *Daphnia* biology, making it is not immediately obvious how a perturbation in a given parameter might influence population performance. Some insight into the direction and magnitude of the effect of a small change in a parameter can be gained from sensitivity analysis. Partial derivatives $\partial z/\partial p$ of a population property z with respect to a parameter p were computed as central, finite difference approximations $\partial z/\partial p \approx \frac{1}{2} \Delta p^{-1} [z(p + \Delta p) - z(p - \Delta p)]$, where Δp is a small perturbation to the nominal parameter value. As parameters differ in both units and magnitude, the relative importance of a parameter cannot be inferred from the partial derivatives alone. Results of the sensitivity analysis are therefore reported as relative parameter sensitivities, which can be interpreted as the exponent of a power function fitted to a small interval containing the nominal parameter value. The relative sensitivity, which by economists is usually termed the elasticity of a function z with respect to a parameter p, is defined as $(p/z)\partial z/\partial p = \partial(\ln z)/\partial(\ln p)]$. A relative sensitivity of 1 implies direct proportionality, while -1 implies inverse proportionality. Relative sensitivities different from unity imply sub- or superlinear proportionality (for example, a relative sensitivity of 2 means that a 1% change in the input parameter gives 2% change in the output value).

By inspection of Fig. 4.21, it is apparent that the two parameters of greatest overall importance to the the performance of a *Daphnia* population are the maximal ingestion rate in neonates (I'_1) and the assimilation efficiency (ε). The equality of relative sensitivities for these two parameters probably reflects that they influence all population properties as the product $\varepsilon\,I'_1$, i.e., the gross assimilation rate. Although not shown in Fig. 4.21, we can expect that food elemental composition, as discussed in Section 4.5, will also have a strong influence on population performance through the constraints placed on food assimilation efficiency by the maintenance of animal elemental composition.

The size dependency of food ingestion, determined by the parameter B'_∞, seems to be of comparatively minor importance. The relative sensitivities to the maintenance costs of respiration and molting (r and h) are closely correlated, with sensitivities to both parameters roughly proportional to their relative magnitudes (5:1). As expected, the incipient limiting food concentration (C') influences only the threshold level for population growth with relative sensitivity of 1; that is, C''_λ is directly proportional to C'.

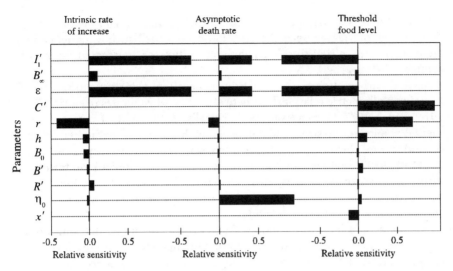

Fig. 4.21. Relative parameter sensitivities of asymptotic population properties: maximum intrinsic rate of increase (λ'), specific death rate (δ), and threshold food level for positive population growth (C''_2). Parameter symbols: I'_1 maximal neonate ingestion rate; B'_∞ extrapolated body size of zero ingestion; ε assimilation efficiency; r respirational loss rate; h molting loss rate; B_0 egg size; B' size of first reproductive investment; R' maximal reproductive investment; η_0 neonate mortality rate; x' age of minimum mortality

The typical life-history parameters B_0 – egg size, B'– size at first reproductive investment, and R' – maximum fraction of net assimilate allocated to reproduction, all have a surprisingly low influence on the asymptotic population properties. The same applies to the parameters of the survival function (η_0 – neonate mortality rate, and x' – age of minimum mortality), with the exception of the asymptotic death rate, which is directly proportional to neonate mortality (η_0).

Within the limitations of the chosen set of performance measures, it appears that the present model is more sensitive to changes in parameters associated with the capture and utilization of food than with the parameters of the individual reproduction and mortality schedule (perhaps rather disappointing, since so much of this chapter has been devoted to life-history events in *Daphnia*). That the model appears to be less sensitive to the age- or size-dependent parameters can be taken as an argument for approximating zooplankton dynamics by an unstructured population model. The validity of such an approximation with age- and size-independent vital rates will probably be highest close to an equilibrium situation where $\lambda \approx 0$ (and $C \approx C''$), and will be weakened by increasing departure from equilibrium.

Asymptotic Population Properties in Other Major Zooplankton Taxa. Throughout this chapter we have used large *Daphnia* as the target for model development. The large amount of available data on a single species

(*D. pulex*) makes it possible to construct a model with a high level of internal consistency, but leaves open the question of how well such a model can be adapted or generalized to other zooplankton taxa or aggregated functional groups.

Several earlier attempts to construct general zooplankton models relied heavily on allometric relationships between size and vital rates (e.g., Hall et al. 1976; Lynch 1977). When examined critically in the light of recent experimental results, many size trends observed within a major taxon, like rotifers or cladocerans, break down when comparisons are made across major taxa (DeMott 1989). There is more uniqueness to common zooplankton taxa than is revealed by size alone; thus, most likely other structural or functional characters will be more instrumental in capturing the essential features of a given zooplankton taxon in a simple model.

From a simple life-history model, Allan (1976) concluded that one would expect the intrinsic rates of increase among the major zooplankton taxa to be ranked as rotifers > cladocerans > copepods (shaded areas in Fig. 4.22). Since then, many common zooplankton species have been successfully cultured and studied in life-table experiments (Table A10.7 provides a partial list of λ' values estimated from a range of species). Although the frequency distribution in Fig. 4.22 shows some resemblance with the predicted pattern, the apparent variability within the major taxa is much higher than predicted by Allan (1976). While the highest values of λ' are generally found in rotifers, and the lowest in the few copepod species that have been investigated, there are also notable exceptions: both the rotifer *Keratella cochlearis* and the cladoceran *Chydorus sphaericus* have maximal intrinsic rates of increase comparable to those of copepods. The maximal intrinsic rate of increase predicted by the present model (0.38 day^{-1}) is seen to be quite close to the median of the frequency distribution in Fig. 4.22 (0.36 day^{-1}).

In a situation where two zooplankton species are involved in exploitative competition for a common food resource, the population with the lowest threshold food level will eventually depress the food resource to a level where the other population is unable to maintain positive net population growth. The threshold food level is therefore thought to be a key parameter in determining the competitive ability of a zooplankton species (e.g., DeMott 1989). In a collection of published threshold food levels from several zooplankton species (Table A10.8), the interspecific variability extends over almost two orders of magnitude, with a frequency distribution resembling the lognormal (Fig. 4.23). The threshold food level predicted by the present model [0.073 (mg C) l^{-1}] is seen to be quite close to the median of the frequency distribution in Fig. 4.23 [0.075 (mg C) l^{-1}].

Although available data are limited, copepods appear to have the lowest threshold food levels, rotifers the highest, while cladocerans are in an intermediate position. The low threshold food level found in the rotifer *Keratella cochlearis* and the high value found in the marine copepod *Acartia tonsa* are both exceptions to this pattern. Still, there appears to be a tendency for

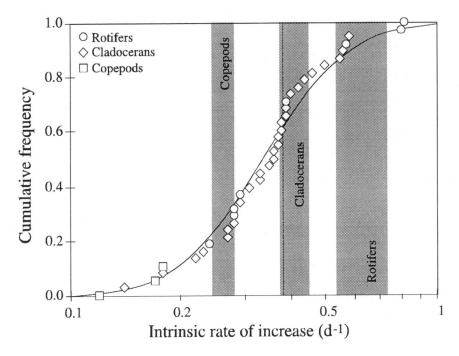

Fig. 4.22. Cumulative frequency distribution of published maximal intrinsic rates of increase from different zooplankton species (Table A10.7); *solid line* is a fitted lognormal distribution; *broken vertical line* is the value corresponding to the *Daphnia* model; *shaded areas* marked *Rotifers, Cladocerans,* and *Copepods* are the characteristic ranges for intrinsic rates of increase in these groups at 20 °C, according to the model of Allan (1976)

a classical *r*- and *K*-selection tradeoff (Pianka 1970) among the investigated zooplankton species; no species has both a low threshold food level, which should be a competitive advantage at limiting food levels, and a high maximal growth rate, which should be a competitive advantage at sudden bursts of high food availability. As a group, rotifers appear to be dominated by *r*-strategists, while copepods have more of a *K*-strategist character. From their intermediate position, it is suggested that cladocerans as a group have a balanced tradeoff between the two competitive traits.

If the median is interpreted as the typical value for the ensemble of zooplankton species presented in Tables A10.7 and A10.8, one might say that the present model describes a typical zooplankton species in terms of both the intrinsic rate of increase and the threshold food level for positive populations growth.

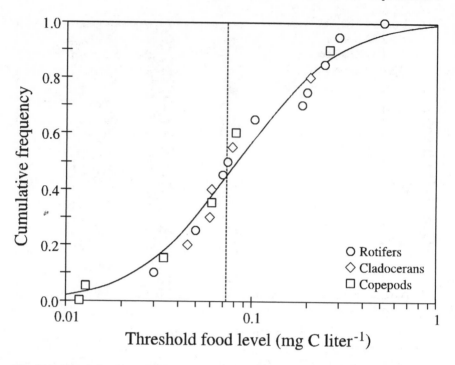

Fig. 4.23. Cumulative frequency distribution of published threshold food levels for positive net population growth from different zooplankton species (Table A10.8); *solid line* is a fitted log-normal distribution; *broken vertical line* is the value corresponding to the *Daphnia* model

4.8 Summary and Conclusions

In the course of this chapter we have formulated a sequence of submodels describing various processes involved in the proliferation of herbivorous zooplankton populations. In contrast to the general approach used in the parametrization of the processes controlling growth and nutrient utilization in phytoplankton populations (Chap. 3), we have throughout this chapter directed the model development toward a small and quite well-defined group of target species that might be collectively referred to as large daphnids (*D. longispina, D. hyalina, D. pulex,* etc.).

This choice might be justified from observing that large daphnids appear to be located close to the medians of the frequency distributions of a least some key population parameters; i.e., the typical zooplankton species should have a certain likeness to this group. In the particular context of pelagic nutrient cycling and lake restoration by biomanipulation, large *Daphnia* species have repeatedly been identified as key species in structuring the phytoplankton communities. For the purpose of model construction, an emphasis on large daphnids has the additional advantage that the large amount of available data on a single species (*D. pulex*) makes it possible to construct a model with a level of internal consistency that probably would have been unattainable for most other common zooplankton species.

Based on a review of the major features of *Daphnia* biology, a model describing the growth and reproduction of *Daphnia pulex* under food-sufficient conditions has been formulated in Section 2 of this chapter. A parametrization based on *D. pulex* data from Lynch (1989) suggests a net assimilation rate linearly decreasing with body size, and that both the transition to maturity and the fraction of net assimilate allocated to reproduction are size-dependent. Simulation experiments point to the importance of molting losses in determining the form of the *Daphnia* body growth curve; as molting losses are taken from the fraction allocated to body growth, body growth is halted at a size where egg production is still high. This effect appears to be responsible for the nearly constant production rate in adult *Daphnia*, as observed by Lynch et al. (1986). The model initially describes the basic events of brood production and molting as explicit instar transitions; a simplified continuous approximation without instar transitions is shown to give a good representation of both body growth and egg production rate in *Daphnia pulex*.

From a model of the allocation of net assimilate into growth, reproduction, and molting under food-sufficient conditions, the corresponding total food intake will be given by accounting for the material lost as egesta and respiration. In Section 3 of this chapter evidence is presented that both respiration and assimilation efficiency can be considered size-independent in *Daphnia pulex*. As net assimilation rate is assumed to be size-dependent; this implies that the saturated ingestion rate must be linearly decreasing with body size. Comparing simulation results with body growth curves for *D. pulex* under food limitation (Taylor 1985) indicates that the incipient limiting food level can also be considered size-independent. With the saturated ingestion rate being size-dependent, this means that the threshold food level for positive body growth should increase with body size, and that there should be a critical food level where body growth is halted before reaching the size of maturity.

In Section 5 the implications of maintaining constant elemental ratios in grazers are investigated. It is shown that while one can construct infinitely many allocation strategies compatible with the constraints from maintaining constant elemental ratios under variable food composition, all these strategies are bounded within the two contrasting extremes with respect to nutrient conservation. The most wasteful strategy would be for the animals to have a constant nutrient excretion rate, independent of food nutrient content. For this strategy there must be a threshold food elemental composition such that positive grazer growth is impossible below this level. At the other extreme, the animals could be able to reutilize metabolic wastes to such an extent that the maximal growth rate is maintained whenever this is compatible with a non-negative nutrient excretion rate. For this strategy there will be no threshold food elemental composition for positive grazer growth; thus positive net assimilation will be possible for all food sources with nonzero nutrient content. From analysis of a data set on *Daphnia* P release (Olsen et al. 1986b), it appears that the phosphorus utilization strategy of *Daphnia* is intermediate between these two extremes. For this intermediate strategy, the net assimilation rate will be directly proportional to food P content, as long as the food has a lower P content than the grazers. This leads to a particularly simple formulation of C- and P-limited growth in *Daphnia*.

Thus far, only processes at the individual level have been considered; when proceeding to the population level, population mortality losses must also be taken into account. In Appendix A4 it is shown that the small cohort sizes used in most life-table studies will introduce a strong stochastic element into the observed survival curves, which would tend to mask out differences between mortality schedules. In Section 6 it is shown that different sets of *Daphnia pulex* survival data obtained under food-sufficient conditions generally agree to within the expected level of experimental error variance. The survival curves generally conform to a convex mortality schedule with high mortality in young and old individuals, and a minimum at an intermediate age.

Practically all experiments on survival in food-limited zooplankton populations have been performed as transfer-culture studies. In Section 6 of this chapter it is argued that the interaction between acute starvation and yolk allocation in the transfer culture regime might create strong maternal effects on offspring survival. A set of *Daphnia pulex* survival curves obtained under varying degrees of food limitation (Frank et al. 1957) could be represented as two distinct subcohorts of eggs; one with the normal mortality schedule of a food-sufficient population, the other with a fixed, high mortality rate, as would be expected for individuals developing from yolk-deficient eggs. This suggests that the immediate effects of food limitation on mortality are limited, at least as long as the animals are able to maintain positive net growth, and that increased juvenile mortality in food-limited transfer cultures is mostly an artifact caused by acute maternal starvation at the end of the brood cycle.

In Section 7 of this chapter, the key processes that have been described in the preceding sections: growth, reproduction and mortality, are assembled into a model of *Daphnia* demography. The model properties are investigated in terms of the asymptotic population structure resulting from unconstrained growth at constant food concentration. The stable age distribution is shown to be dominated by eggs and early juveniles in all populations with positive net growth rate. When the net population growth rate is decomposed into population-specific, asymptotic birth and death rates, it is shown that the death rate is close to 5% of the maximal birth rate for all food levels supporting positive population growth.

A parameter sensitivity analysis revealed that all the investigated population properties were most sensitive to parameters associated with the efficiency of food collection and food utilization. By comparison, typical life-history parameters, including egg/neonate size, size at first reproductive investment and maximum reproductive investment, had only minor influence on population performance. The sensitivities to size- or age-dependent parameter variations were small for most population performance measures. This might be taken as an argument for using a simple, unstructured population model as an approximation to the dynamics of *Daphnia* populations under certain circumstances.

4.9 Symbols, Definitions, and Units

Symbol	Definition	Dimension
α	Fraction of cohort developing from aberrant eggs	–
β	Asymptotic, population specific birth rate	day^{-1}
B	Body size	$\mu g\ C$
B_0	Egg size	$\mu g\ C$
B_1	Neonate body size	$\mu g\ C$
B_i	Body size at the start of instar i	$\mu g\ C$
B'	Body size at first reproductive investment	$\mu g\ C$
B''	Body size at first maternity	$\mu g\ C$
B'''	Maximum body size	$\mu g\ C$
B_∞	Extrapolated body size where net assimilation is zero	$\mu g\ C$
B'_∞	Extrapolated body size where ingestion is zero	$\mu g\ C$
C	Food carbon concentration	$(mg\ C)\ l^{-1}$
C'	Incipient limiting food concentration	$(mg\ C)\ l^{-1}$
C''	Threshold food level for positive individual growth	$(mg\ C)\ l^{-1}$
C'_1	Threshold food level for positive neonate growth	$(mg\ C)\ l^{-1}$
C''_R	Threshold food level for reproduction	$(mg\ C)\ l^{-1}$
C''_λ	Threshold food level for positive netpopulation growth	$(mg\ C)\ l^{-1}$
D_i	Duration of instar i	days

Symbol	Definition	Dimension
δ	Asymptotic, population specific death rate	day^{-1}
δ'	Specific death rate in a food-sufficient population	day^{-1}
ε	Food carbon assimilation efficiency	–
$\overset{\bullet}{\varepsilon}$	Maximum food carbon assimilation efficiency	–
E	Accumulated egg carbon production	$\mu g\ C$
E_i	Egg carbon production during instar i	$\mu g\ C$
g	Specific net assimilation rate	day^{-1}
g_1	Specific net assimilation rate in neonates	day^{-1}
g_i	Specific net assimilation rate during instar i	day^{-1}
g'	Net assimilation rate on optimally composed food	day^{-1}
h	Specific molting loss rate	day^{-1}
I	Specific food ingestion rate	day^{-1}
I'	Maximum specific ingestion rate at saturating food	day^{-1}
I'_1	Maximum specific ingestion rate in neonates	day^{-1}
K^{*}_2	Maximum net growth efficiency	–
$l(x)$	Fraction of individuals surviving to age x (survival function)	–
$l'(x)$	Survival function for aberrant individuals	–
$\bar{l}\,(x)$	Composite survival function	–
λ	Intrinsic rate of population increase (asymptotic net growth rate)	day^{-1}
λ'	Maximum intrinsic rate of population increase	day^{-1}
M_i	Body mass lost through molting at the end of instar i	$\mu g\ C$
$m\,(x)$	Maternity rate (maternity function)	day^{-1}
η	Instantaneous mortality rate (force of mortality)	day^{-1}
η_0	Instantaneous mortality rate at age 0	day^{-1}
η'	Instantaneous mortality rate in aberrant individuals	day^{-1}
θ	P content of animals	$(\mu g\ P)\ (mg\ C)^{-1}$

Symbol	Definition	Dimension
Q	P content of food particles	$(\mu g\,P)\,(mg\,C)^{-1}$
P	Food P concentration	$(\mu g\,P)\,l^{-1}$
ρ	Specific P release rate	$(\mu g\,P)\,(mg\,C)^{-1}\,day^{-1}$
R	Fraction of net assimilate allocated to reproduction	–
R_i	Fraction allocated to reproduction during instar i	–
R'	Maximum fraction allocated to reproduction	–
r	Specific respiration rate	day^{-1}
x	Individual or cohort age	days
x'	Age of minimum mortality	days

5 Nutrients, Algae and Herbivores - the Paradox of Enrichment Revisited

Rosenzweig's results might more reasonably have been used to prompt questions such as the following: What are the critical values of enrichment? How does the time to extinction of a system vary with the degree of enrichment? How do critical levels of enrichment and time to extinction vary with other parameters? Why does nature not collapse?
McAllister, LeBrasseur and Parsons (1972).

Lotka (1925) and Volterra (1926) independently developed simple mathematical descriptions of the interaction between predators and their prey. These theories later became the ancestors of a whole family of what is collectively known as Lotka-Volterra models. Elementary ecology text books usually concentrate on the simplest kind of Lotka-Volterra model, where the prey have an unbounded capacity for exponential growth in the absence of predators, and where the predators have an unbounded capacity for killing prey. As noted by May (1975), such a system has certain pathological dynamic properties which are equivalent to the neutral stability of a frictionless pendulum: the system oscillates forever with an amplitude that is determined solely by the initial conditions.

In more realistic models where either prey growth is bounded by a finite carrying capacity, or where predator capacity for capturing prey saturates at a finite level of prey abundance, the neutral stability property is lost (that is, the neutral stability is structurally unstable; May 1975). For two-dimensional Lotka-Volterra systems with one predator and one prey species, the powerful Poincaré-Bendixson theorem can be used to prove that essentially all such systems will either settle down to a stable steady state, or end up in a limit cycle with both the oscillation period and amplitude being independent of the initial conditions (May 1975).

That two-dimensional prey-predator systems exhibit only stable equilibrium points or stable cycles is related to the fact that a closed orbit in the plane has a defined inside and outside. When stepping up to three or more dimensions, one can no longer distinguish the inside from the outside of a closed orbit, which again implies that no equivalent of the Poincaré-Bendixson theorem exists in higher than two dimensions. Three-dimensional prey-predator systems can therefore display a full and rich dynamic complexity, including strange attractors and chaos (May 1981).

Whether a two-dimensional Lotka-Volterra system is attracted to a stable equilibrium or to a limit cycle depends on the details of the predator-prey interaction. Stability studies have shown that the presence of a carrying capacity for the prey population tends to dampen the oscillations, while the presence of a saturating predator response tends to destabilize the system. Rosenzweig (1971) has shown by graphical arguments that when the prey carrying capacity is increased, the stabilizing effect of prey density dependence will weaken. The result is the famous paradox of enrichment which states that a prey-predator system can be driven from a stable equilibrium to a qualitatively different limit cycling behavior by an enrichment affecting the potential yield of prey in the system.

Rosenzweig (1971) interpreted this result as a warning that "man must be careful in attempting to enrich ecosystems in order to increase its food yield. There is a real chance that such activity may result in a decimation of the food species that are wanted in greater abundance." From their experiences with whole-lake fertilization experiments made in order to increase fish yield, McAllister et al. (1972) found Rosenzweig's interpretation of the paradox of enrichment to be overly pessimistic, and suggested that future research should be directed toward how the possible existence of critical levels of enrichment could be utilized in practical ecosystem management. Apparently, little progress has been made in this direction since then.

After the introduction of the concept of lake restoration by biomanipulation (Shapiro et al. 1975), a variant of the paradox of enrichment turns up in a new context. The purpose of a biomanipulation is to improve water quality by reducing the predation pressure on the zooplankton so that a stable equilibrium with grazer-controlled algal biomass is established. Benndorf (1987) observed that in the lakes where successful biomanipulation had been reported, the nutrient loading was either moderate or unknown, and that in the few highly eutrophic lakes where biomanipulation had been attempted, it apparently had failed to improve water quality. This observation has a clear resemblance to the paradox of enrichment: it seems that when the carrying capacity of the algae is above some critical level, in terms of nutrient loading, the establishment of a stable steady state with grazer-controlled algal biomass becomes unlikely.

The total amount of nutrients available for plankton growth is determined by the balance between nutrient inputs to and nutrient losses from the pelagic zone. The nutrient loss processes are to a large extent mediated by the plankton organisms themselves through sinking and mortality, as discussed in Chapter 2. As all members of the plankton need essential nutrients to grow, the net amount of nutrients available for algal growth is determined by the partitioning of nutrients among the community members. The actual yield of algal biomass at a given level of algal-bound nutrient depends on the growth rate of the algae, as discussed in Chapter 3. This means that the algal carrying capacity in terms of the potential yield of algal biomass becomes a dynamic entity related to both the nutrient supply rate and the processes of nutrient utilization and recycling in the plankton community. The carrying capacity can therefore not be considered an external forcing variable, as is done in most prey-predator models.

This has the important consequence that a minimal model of the interaction between algae and grazers in nutrient-limited plankton communities needs to consist of at least three state variables: the biomasses of phyto- and zooplankton and the algal carrying capacity, either in terms of total available nutrient or as cellular nutrient content of the algae. Although the inclusion of an additional state variable might seem a minor extension to the two-species Lotka-Volterra models, it can, in fact, represent a major qualitative change in dynamical complexity (May 1975).

5.1 Nutrients, Algae, and Grazers - a Minimal Model

Let us consider, as the simplest possible case, a plankton community consisting of a single algal prey species and a single zooplankton grazer species. The nutrient input to the lake is assumed to be such that phosphorus is the only mineral nutrient likely to become limiting to algal growth. Algal growth is assumed to be controlled by the cellular P content, as outlined in Section 3.1. The dynamics of the algal population is determined by the balance between growth and loss, where losses are comprised by dilution, sinking, and grazing. The algal species is assumed to be edible and otherwise nutritionally suitable for grazer growth and reproduction. We can think of cryptomonads as typical representatives of this kind of phytoplankton community.

Grazer feeding and assimilation is assumed to be size- and age-independent, such that grazer growth can be described by an unstructured population model. Net growth rate is assumed to be determined by both algal biomass and algal phosphorus content as outlined in Section 4.5 in the previous chapter. The dynamics of the grazer population is determined by the balance between growth and loss, where losses are composed of dilution and mortality. Higher trophic levels are assumed to be absent or negligible, so that grazer mortality is exclusively nonpredatory. By disregarding the demographic aspects of grazer population dynamics, we will have to assume a constant specific mortality rate corresponding to the asymptotic death rate of the stable age distribution.

Mass-Balance Equations. In Section 2.5 we formulated a general description of the phosphorus cycle in a plankton community with an arbitrary number of primary producers and consumers. With only one phytoplankton and one zooplankton population, we can drop all indices in the formalism of Section 2.5, so that the biomasses of algae and grazers are represented by the state variables C and Z [both in units of $(mg\ C)\ l^{-1}$].

Noticing that the specific ingestion rate I (day^{-1}) of the grazer is related to the specific clearance rate F [l (mg C)$^{-1}$ day^{-1}] by $I = F\,C$, we can reexpress Eqs. (2.12) and (2.8), the mass-balance equations for phytoplankton and zooplankton biomasses, as

$$\dot{Z} = \left(g - (\delta + D)\right)Z. \tag{5.1}$$

$$\dot{C} = \left(\mu - (\sigma + D)\right)C - I\,Z \tag{5.2}$$

The leading term in Eq. (5.1) is simply the net growth rate of the grazer population, while the first term in Eq. (5.2) is the net phytoplankton growth rate in the absence of grazers, and the second term in Eq. (5.2) is the contribution from grazing losses. μ and g are the specific growth rates (day^{-1}) of algae and grazers, respectively, while σ and δ are specific loss rates (day^{-1}) due to phytoplankton sedimentation and zooplankton mortality. Both algae and grazers suffer losses from dilution, determined by the dilution rate D (day^{-1}).

In Section 2.5, the assumption of homeostatic control of zooplankton elemental composition is utilized to make an implicit representation of the fraction of total P allocated to zooplankton. The remainder of total available P is partitioned between the external pool of inorganic P and the internal pools of cellular P in the competing phytoplankton species. While the pool of inorganic P will generally be small, it is still instrumental in representing algal competition for phosphorus. In a system with a single phytoplankton species there is no need for an explicit representation of resource competition, so that the pools of inorganic and cellular phosphorus can safely be aggregated into a single pool of algal P (or, more precisely, potentially available algal P). Adding Eqs. (2.7) and (2.13), describing the dynamics of cellular and inorganic P, and observing that the terms involving zooplankton ingestion cancel out, the mass-balance of algal phosphorus [P, (μg P) l^{-1}] becomes

$$\dot{P} = D\,P_{I.} - (\sigma + D)P - g\,\theta\,Z. \tag{5.3}$$

The first term of Eq. (5.3) means that the external supply of P to the system enters directly into algal P, while the second term says that algal P is lost by sinking and flushing at the same rate as algal biomass. The last term of Eq. (5.3) means that only the fraction of P incorporated into new grazer biomass is lost from the pool of algal P by zooplankton grazing, the remainder of ingested P is recycled and immediately returned to the algae.

Rate Processes and Process Parameters. In order to make the model analytically tractable, we will have to assume that seasonal variation in light and temperature can be neglected, such that all model parameters are constant over time.

If we assume that algal growth is a process controlled only by phosphorus availability, this can conveniently be described by the Droop model [Eq. (3.2)], which with the present choice of state variables can be expressed as

$$\mu = \mu'\left(1 - Q'\frac{C}{P}\right). \tag{5.4}$$

The medians from the frequency distributions in Figs. 3.2 and 3.3 are probably the best estimates we can make for typical parameter values in Eq. (5.4). From this, we can assume a phosphorus subsistence quota $Q' = 3.8$ (μg P) (mg C)$^{-1}$ and a maximal growth rate $\mu'' = 1.2$ day^{-1}. As a consequence of the approximations made in the model equations (5.3), (5.4), algal P quota is not constrained below an upper limit Q'', as in Eq. (3.3). Allowing the algae to have infinite P storage capacity means that the asymptote of the Droop model will be equal to the maximal growth rate ($\mu' = \mu''$). As neither algae nor grazers are likely to be phosphorus-limited when the P quota exceeds Q'', the chosen representation should have only minor consequences for the performance of the model.

While large, nonmotile plankton algae are likely to have sinking loss rates that are at least an order of magnitude higher than observed net loss rates of phosphorus in lakes, flagellates with active vertical movement should probably suffer only negligible losses from sinking. As a compromise between these two extremes, one can assume the algae to have a sinking loss rate equal to average net loss rate of phosphorus, as estimated for the retention model [Eq. (2.5); that is, $\sigma = 0.0008$ d^{-1}]. This choice of parameter value will make the P retention in an ungrazed system equal to the observed average for lakes with the same dilution rate, and at the same time ensure that phytoplankton sinking will constitute only a minor P flux under most circumstances.

If we want the unstructured grazer model to conform as closely as possible to the age-structured *Daphnia* model of Chapter 4, we can require that the grazer population must have a net growth rate less than or equal to the maximal intrinsic rate of increase (λ') determined in Section 4.7, that is, $g - \sigma \leq \lambda'$. Assuming that $\lambda' = 0.38$ day^{-1} and that $\delta = 0.02$ day^{-1} gives a maximal zooplankton growth rate $g' = 0.4$ day^{-1}. If ε is the assimilated fraction of ingested food carbon, I' is the saturated ingestion rate (day^{-1}), and r is the specific rate of respiration (day^{-1}), we can express the zooplankton growth rate under food-sufficient conditions as the difference between assimilation and respiration, or as $g = \varepsilon I' - r$. If we assume that ε and r are equal to the size-independent parameter values determined in Section 4.3 (that is, $\varepsilon = 0.8$ and $r = 0.25$ day^{-1}), the maximal ingestion rate will be given by $I' = (g' + r)/\varepsilon = 0.81$ day^{-1}.

If we use the piecewise linear functional response [Eq. (4.9)], the specific ingestion rate I (day^{-1}) as function of food concentration can be written as

$$I = I'\text{Min}(1, C/C'), \tag{5.5}$$

where C' is the incipient limiting food level $[(\text{mg C}) \, l^{-1}]$. If we require that the unstructured grazer model should have the same threshold food level for positive net population growth as determined in Section 4.7 $[C''_\lambda = 0.0725$ $(\text{mg C}) \, l^{-1}]$, we must have $g = \delta$ when $C = C''_\lambda$, or $C' = \varepsilon I' C''_\lambda / (r + \delta) = 0.17$ $(\text{mg C}) \, l^{-1}$.

From the discussion of the phosphorus economy in *Daphnia* in Section 4.5, it was concluded that phosphorus-limited growth could be described by a piecewise linear function of food P content. This can be expressed in terms of the present state variables as

$$g = (\varepsilon I - r)\text{Min}(1, P/\theta C), \tag{5.6}$$

where θ is the grazer P content [which is found to be close to 30 $(\mu\text{g P}) \, (\text{mg C})^{-1}$ in *Daphnia* species, and generally lower in other zooplankton taxa].

The last two model parameters, the dilution rate $(D; \text{day}^{-1})$ and the input P concentration $[P_L; (\mu\text{g P}) \, l^{-1}]$, are considered external forcing variables much in the same way as in total phosphorus loading models. The dilution rate enter the mass-balance equations both as the product DP_L, which is the volumetric P load to the pelagic zone, and as the loss rate due to flushing. The model parameter P_L will be less than the measured, flow-weighted input concentration to a given lake if some of the load is lost before entering the pelagic zone. If we assume an average 22% load decay, as suggested by Fig. 2.2, P_L will be 78% of the total external P load.

5.2 Equilibrium Points and Local Stability

Every combination of the three state variables P, C, and Z will correspond to a point in the state space of the system [Eqs. (5.1)-(5.3)]. As all state variables represent physical entities, they must necessarily all be non-negative; thus, the state space is confined to the positive cone ($P \geq 0$, $C \geq 0$, $Z \geq 0$). At certain points in state space, which are called the stationary points of the system, all the right-hand sides in Eqs. (5.1)-(5.3) will evaluate to zero (that is, $\dot{P} = \dot{C} = \dot{Z} = 0$). If the system is initially located exactly at such a stationary point, it will remain there indefinitely, unless it is driven away from it by some external disturbance. If the system returns to the stationary point after some small perturbation, the stationary point is said to be locally stable.

In Appendices A6 and A7 it is shown that the differential equation system (5.1)-(5.3) possesses several stationary points, whose existence and stability properties depend on the phosphorus loading and water renewal conditions in terms of the input P concentration P_L and the dilution rate D. Two of the stationary points, which in Appendix A6 are called the washout point and the grazer extinction point, are located on the boundary of the positive cone. The washout point, corresponding to the absence of both algae and grazers, is in Appendix A7 shown to be locally unstable for all dilution rates $D < \mu' - \sigma$. This means that a sterile system can be invaded from an arbitrary small inoculum of living organisms, as long as the dilution losses do not exceed the maximal net growth rate of the algae.

Persistence and the Grazer Extinction Point. The grazer extinction point is characterized by zero zooplankton biomass and nonzero phytoplankton biomass (that is, $Z = 0$, $C > 0$). In the absence of grazing, the algal growth rate will end up at equilibrium with sinking and dilution losses ($\mu = \sigma + D$). It is shown in Appendix A7 that this stationary point will be locally unstable if the grazers are able to maintain positive net growth rate when feeding on algae growing at a specific rate $\mu = \sigma + D$, and locally stable if not. This result is in accordance with common intuition if it is expressed as: grazers will only be able to invade an algal community at equilibrium with dilution and sedimentation losses if they are able to maintain positive net population growth on this food resource.

An ecological system is said to be persistent if all state variables with positive initial conditions remain nonzero over time. Gard and Hallam (1979) demonstrated that in prey-predator systems, persistence of the top predator is equivalent to entire system persistence. From this, we can infer that the system will be persistent if it is repelled from the grazer extinction point; that is, if this point is locally unstable. In Appendix A8 it is shown that this stability condition can be expressed in terms of a critical dilution rate D'_-, called the persistence boundary, so that the system will be bounded away from the grazer extinction point for dilution rates $D < D'_-$. Substituting the model parameters in Table 5.1 into Eq. (A8.4) gives $D'_- = 0.032$ day^{-1}, corresponding to a critical water residence time around 1 month. In the data set compiled by Prairie (1988; Fig. 2.2), 70% of the lakes have average dilution rates below the persistence boundary, although this data set can probably not

Table 5.1. Summary of model parameters

	Parameter	Value	
μ'	Maximum algal growth rate	1.2	day^{-1}
σ	Algal sinking loss rate	0.0008	day^{-1}
Q'	Minimum algal P content	3.8	$(\mu g\ P)\ (mg\ C)^{-1}$
ε	Maximum assimilation efficiency	0.8	–
I'	Maximum ingestion rate	0.81	day^{-1}
C'	Incipient limiting food concentration	0.17	$(mg\ C)\ l^{-1}$
r	Respiration rate	0.25	day^{-1}
δ	Grazer mortality loss rate	0.02	day^{-1}
θ	Grazer P content	30.0	$(\mu g\ P)\ (mg\ C)^{-1}$

be considered a representative random sample of lakes in general; as the data were collected to investigate the relationship between flushing and retention, it is very likely that lakes with very low and very high dilution rates are over-represented.

By evaluating the relative parameter sensitivities of the persistence boundary [Eqs. (A8.7)-(A8.10)] for the parameter values in Table 5.1, we find that D'_- is very robust to changes in the algal growth and loss parameters: a 1% increase in the maximal algal growth rate (μ') or the algal sinking loss rate (σ) results in only a 0.05% decrease or a 0.001% increase in D'_-. On the other hand, the persistence boundary is very sensitive to changes in grazer growth and loss parameters: a 1% increase in the maximal grazer growth rate (g') or the grazer mortality rate (δ) results in either a 1.7% increase or a 0.7% decrease in the critical dilution rate D'_-. The result suggests that a grazer population should be most vulnerable to predation when the water renewal time is close to the persistence boundary; in that case, only a modest increase in the mortality rate (δ) caused by predatory losses might be sufficient to make the grazer extinction point stable.

The persistence boundary is also very sensitive to changes in the stoichiometric properties of the organisms: a 1% increase in the algal subsistence quota (Q') or the grazer P content (θ) results in either a 1.7% increase or a 1.7% decrease in the critical dilution rate D'_-. This result may suggest that the rotifers should be the zooplankton group most likely to succeed in lakes with high dilution rates, as many rotifers have high maximal growth rates (cf. Fig. 4.22) and low phosphorus requirements compared to other groups (Hessen and Lyche 1991).

For dilution rates below the persistence boundary, grazer extinction can also result if the P loading is sufficiently low. If the input concentration is below a critical level P'_L, defined by (A6.23), there will not be any stationary points with non-zero zooplankton biomass. Substituting the model parameters in Table 5.1 into Eq. (A6.23) gives $P'_L = 0.68$ (μg P) l^{-1} for $D = D'_-$. With the lowest total P value observed in the NIVA survey shown in Fig. 2.3 being 1.4 (μg P) l^{-1}, we can expect the critical input concentration P'_L to be exceeded in all of the 355 Norwegian lakes investigated in that survey, even if we assume zero P retention and no load decay.

If the input concentration is above the critical level ($P_L > P'_L$), and the dilution rate is above the persistence boundary ($D > D'_-$), an additional internal stationary point comes into existence. This stationary point is characterized by grazer growth rate being limited only by algal P content, and not by algal biomass in terms of carbon. In Appendix A6 it is shown that at this stationary point, algal growth rate is independent of the P loading conditions, while both phyto- and zooplankton biomasses are directly proportional to the input concentration P_L, and thus, also proportional to each other. In Appendix A7 it is shown that if this stationary point exists, then it will always be locally unstable.

The Internal Equilibrium. If the input concentration is above the critical level ($P_L > P'_L$), and the dilution rate is below the persistence boundary ($D < D'_-$), it is shown in Appendix A6 that there will be only one internal stationary point (that is, with all state variables being nonzero). This stationary point is characterized by an algal biomass below the incipient limiting level ($C < C'$), such that zooplankton growth will be food carbon-limited. At input P concentrations below a critical level P''_L, defined by Eq. (A6.24), zooplankton growth will be simultaneously C- and P- limited at the internal stationary point. In Appendix A7 it is shown that the internal stationary point will be unconditionally locally stable, and can thus be called an internal equilibrium.

The analytical expressions developed in Appendix A6 imply that the equilibrium point traces out a continuous curve when the input phosphorus concentration P_L is increased from P'_L and upward. When the steady state biomasses are viewed as functions of P_L (Fig. 5.1), it is seen that C decreases while Z increases with increasing phosphorus enrichment. At the same time, the algal growth rate increases in proportion with grazer

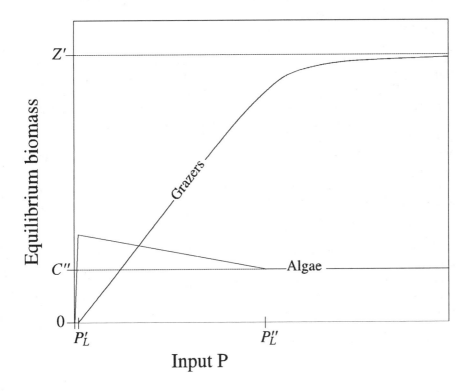

Fig. 5.1. Phyto- and zooplankton biomasses at the internal equilibrium point as function of input phosphorus concentration

biomass from $\sigma + D$ towards the asymptotic maximum μ'; the channeling of increased P loading into zooplankton biomass is accompanied by a corresponding increase in primary production in order to support the resulting level of secondary production. At low phosphorus loading, when grazer growth is limited by both C and P, Eq. (A6.18) implies that algal biomass will be linearly decreasing with increasing grazer biomass. When the input concentration exceeds the critical level P''_L, where the algal P content becomes non-limiting to grazer growth, the algal biomass remains a constant level C'', given by Eq. (A6.19), while the grazer biomass continues to increase towards an asymptotic level Z', given by Eq. (A6.16).

Equations (A6.16) and (A6.19) show that the asymptotic biomass levels of zoo- and phytoplankton (Z' and C'') are both directly proportional to the incipient limiting concentration (C') and inversely proportional to the maximum ingestion rate (I'). In other words, decreasing the food collection efficiency of the grazer, in terms of the maximal clearance rate (I'/C'), would increase the asymptotic biomasses of both algae and grazers. The asymptotic grazer biomass (Z') is directly proportional to the maximal algal net growth rate $[\mu' - (\sigma + D)]$, while the asymptotic algal biomass (C'') is directly proportional to the sum of grazer loss rates ($r + \delta + D$) and inversely proportional to the assimilation efficiency (ε) of the grazer.

As a change in the asymptotic grazer biomass will also change the equilibrium zooplankton level over the whole domain of the internal equilibrium point, we would expect that increasing maximal algal growth rate (μ') or decreasing algal sinking loss rate (σ) should increase the equilibrium zooplankton biomass, without having any effect on algal biomass. Likewise, increasing zooplankton respiration (r) or mortality (δ), or decreasing assimilation efficiency (ε) should increase equilibrium phytoplankton biomass, without having any effect on zooplankton biomass. Such reciprocal relationships between prey parameters and equilibrium predator biomasses, and vice versa, are so common in prey-predator models that the phenomenon is usually referred to in text books by a special name: *the Volterra principle* (e.g., Roughgarden 1979 or May 1981).

Summarizing the local stability analysis, it can be said that if the dilution rate is such that the system is persistent ($D < D'_-$), there will be two locally unstable stationary points both with $Z = 0$ and one locally stable stationary point with $Z > 0$. If $D > D'_-$ the stationary point corresponding to grazer extinction will become locally stable, while an additional locally unstable stationary point corresponding to P-limited grazer growth will come into existence. The local stability of two remaining stationary points will be unaffected by dilution rate crossing the persistence boundary D'_-.

5.3 Isoclines and Global Stability

In a general, two-dimensional dynamic system described by a pair of differential equations $\dot{x}_1 = f_1(x_1, x_2)$ and $\dot{x}_2 = f_2(x_1, x_2)$, it is usually quite easy to study the global stability properties by graphical analysis of the so-called isocline curves; $f_1(x_1, x_2) = 0$ and $f_2(x_1, x_2) = 0$ (e.g., Rosenzweig and

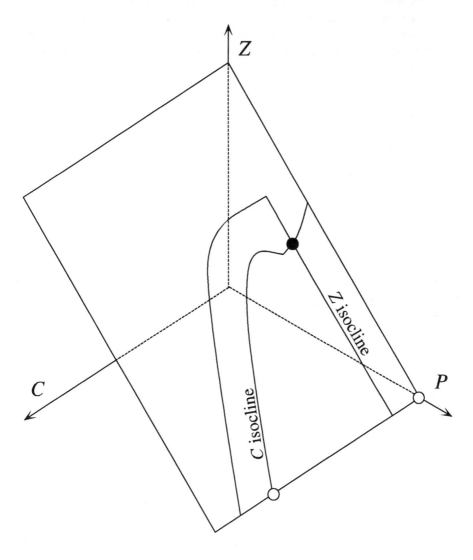

Fig. 5.2. Location of the plane defined by Eq. (5.7) in the (P, C, Z) space, showing intersections with the isocline surfaces for the algae (C) and the grazers (Z). *Open circles* mark locally unstable stationary points; *filled circles* mark locally stable stationary points

MacArthur 1963; Tilman 1980). Canale (1970) showed that in a nutrient-controlled prey-predator system with nutrient conservation, the long-term dynamics will be confined to a plane, and thus reducible to a system with only two state variables. As the present model is irreducibly three-dimensional due to the non-conservative representation of total phosphorus, the isoclines will be surfaces in (P, C, Z) space, which are much harder to visualize and analyze graphically.

The visualization of the zero isoclines is simplified by observing that all the stationary points of the system in Eqs. (5.1)-(5.3) will either have zero net grazer growth rate $(g = \delta + D)$ or zero grazer biomass $(Z = 0)$. This means that all stationary points (which are equivalent to intersections of the isocline surfaces) will be located in the plane generated by substituting $g = \delta + D$ into Eq. (5.3) and setting $\dot{P} = 0$:

$$\left(1 + \frac{\sigma}{D}\right)P + \left(1 + \frac{\delta}{D}\right)\theta Z = P_L. \qquad (5.7)$$

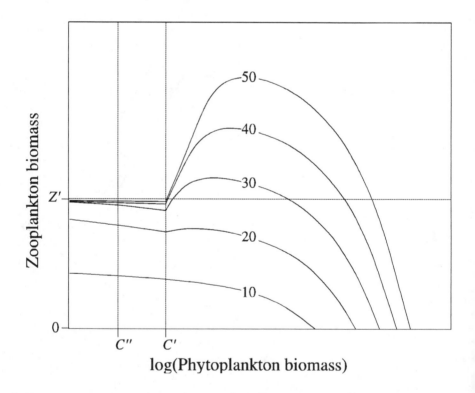

Fig. 5.3. Phytoplankton zero net growth isoclines for different P loading conditions, projected on the (C, Z) plane (logarithmic C axis): *curve labels* are input P concentrations $[P_i; (\mu g\, P)\, l^{-1}]$

The equilibrium plane [Eq. (5.7)] is parallel to the C axis and intersects the P and Z axes at $P = (1 + \sigma/D)^{-1}P_L$ and $Z = (1 + \delta/D)^{-1}\theta^1 P_L$. The intersection between the equilibrium plane [Eq. (5.7)] and the surfaces defined by $\dot{C} = 0$ and $\dot{Z} = 0$ will generate isocline curves in the state space (Fig. 5.2).

The Algal Isocline. The important properties of the isocline curves in Fig. 5.2 are preserved by representing their projection on the (C, Z) plane. The algal isocline, defined by setting $\dot{C} = 0$ in Eq. (5.2), changes its shape with increasing phosphorus loading (Fig. 5.3). At low P supply rates, the algal isocline will be a strictly decreasing function. When the nutrient loading is increased, the discontinuity caused by the piecewise linear functional response [Eq. (5.5)] becomes more prominent. For algal biomass below the incipient limiting food level ($C < C'$) the algal isocline will have a nonpositive slope for all P loading levels; with increasing P loading, the slope of this part of the isocline will increase and eventually become zero when the isocline approaches the asymptotic zooplankton biomass level at the internal equilibrium (Z'). For $C < C'$ the algal isocline develops a segment with positive slope which becomes progressively larger with increasing P loading. Rosenzweig (1969) has pointed out that such a "hump" on the prey isocline will tend to destabilize the interaction between predator and prey.

The Grazer Isocline. Contrary to the algal isocline, the grazer isocline maintains the same general shape with increasing P loading (Fig. 5.4). The grazer isocline is a concave curve with a vertical left flank at the threshold food level for positive population growth (C''). As the total phosphorus content of the system is finite, the P content of the algae will decrease as the algal biomass increases. The decreasing food quality, in terms P content, makes the grazer isocline bend down to the right, meeting the C axis at the point where the algae become so nutritionally poor that they are unsuitable for supporting grazer growth.

The shape of the grazer isocline in the present model is a direct consequence of considering phosphorus to be an essential and potentially limiting element for both algae and grazers, creating an interdependency between the carrying capacities of predators and prey. In models without such stoichiometric constraints on predator growth, the predator isocline will simply either be a vertical straight line through the threshold food concentration for positive predator growth, or a rectangular function branching off to the right as the predator density reaches its carrying capacity, as in Rosenzweig and Mac Arthur (1963).

Freedman and Wolkowicz (1986) have shown that the predator isocline can have a negative slope at high prey densities if the prey has evolved mechanisms of group defence. As an example of group defence they cite that a lone musk ox can be successfully attacked by a pack of wolves, while very few successful attacks are observed on larger herds of, say, six to eight individuals. Freedman and Wolkowicz (1986) show that the presence of a

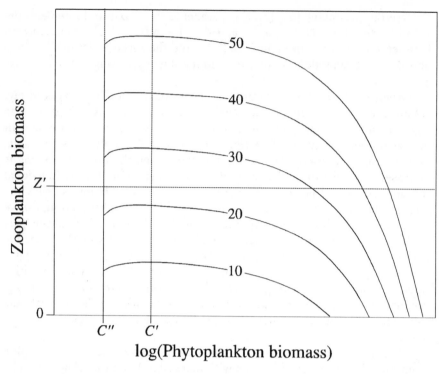

Fig. 5.4. Grazer isoclines for different P loading conditions projected on the (C, Z) plane (logarithmic C axis). *Curve labels* are input P concentrations [P_L; $(\mu g\ P)\ l^{-1}$]

segment with negative slope on the predator isocline has a more dramatic impact on the persistence of a prey-predator system than the presence of a segment with positive slope on the prey isocline, as discussed by Rosenzweig (1969).

While a "hump" on the prey curve can destabilize the equilibrium between predators and prey, the system will still be confined to an asymptotically stable periodic orbit, excluding the possibility of deterministic predator extinction, as pointed out by Gilpin (1972) and May (1972) in their critique of Rosenzweig's (1971) interpretation of the paradox of enrichment. On the other hand, Freedman and Wolkowicz (1986) show that if the predator curve has a "hump", then enrichment can actually lead to the deterministic extinction of the predator and thus lead to the result predicted by Rosenzweig (1971), albeit under somewhat different assumptions.

Freedman and Wolkowicz (1986) state that they are unaware of literature reports of prey-predator models incorporating group defence and similar kinds of consumer inhibition at high prey densities. From the discussion above, it appears that the stoichiometric relationships between phytoplankton algae and zooplankton consumers lead to a kind of accidental or involuntary group defence against predation. As increasing yield per unit nutrient with decreasing growth rate, and other phenotypic adaptations to nutrient limitation, appear to be common to all phytoplankton algae, group defence among plankton algae is probably not a result of coevolution. This means that the presence of group defence-like mechanisms in plankton algae is not dependent on any arguments relying on group selection, and thus not affected by the problem of "cheaters" which is otherwise likely to turn up in the context of group defence and prey-predator coevolution (e.g., Roughgarden 1979).

Isocline Analysis. The algae will have positive net growth at all points located between the algal isocline and the C axis in the equilibrium plane [Eq. (5.7)], and negative net growth at all points outside this region (on the boundary of this region, which is the algal zero net growth isocline, algal growth will be zero). Likewise, the grazers will have positive net growth at all points located between the grazer isocline and the C axis, and negative net growth at all points outside this region. Superimposing the grazer isocline on the algal isocline will partition the equilibrium plane into a disjunct set of regions, each characterized by a combination of signs of the net growth rates of algae and grazers in their interior (Fig. 5.5).

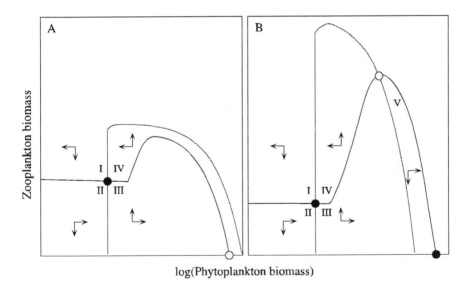

Fig. 5.5A,B. Isocline partitioning of the equilibrium plane [Eq. (5.7)]. *Solid lines* Algal isoclines; *broken lines* grazer isoclines; *vertical arrows* indicate the sign of the net grazer growth rate within a given region; *horizontal arrows* indicate the sign of the net algal growth rate; *filled circles* locally stable stationary points; *open circles* locally unstable stationary points. **A** System with dilution rate below the persistence boundary. **B** System with dilution rate above the persistence boundary

If we consider the case where the dilution rate is below the persistence boundary (Fig. 5.5A), we can follow the trajectory of the system from an initial location in the vicinity of the internal equilibrium, say, somewhere in the region labeled I in Fig. 5.5A. In region I, both algal and grazer growth rates will be negative, so both biomasses will decrease with time. As time evolves, the trajectory will eventually cross the algal isocline and enter region II. In region II, algal growth rate is positive, such that algal biomass will increase with time until the trajectory crosses the grazer isocline and enters region III. In region III, both algal and grazer growth rates will be positive, such that both biomasses will increase with time. This means that the system will eventually leave the region bounded by the algal isocline and enter region IV. In region IV, growth rates have opposite signs, such that algal biomass will decrease while grazer biomass will increase with time, until the system finally enters region I again. We would thus expect the typical system behavior to be counterclockwise orbits around the internal equilibrium. Although local stability analysis indicates that some trajectories should spiral into the internal equilibrium point, the qualitative isocline analysis does not exclude the possibility of some trajectories ending up in a closed periodic orbit (limit cycle).

The analysis for the case where the dilution rate is above the persistence boundary (Fig. 5.5B) will initially be very similar to the previous case. If we initiate the system from the vicinity of the internal equilibrium point in region I, the trajectory will proceed to region III via region II, as before. The main difference from the previous case is that the trajectory can now follow two different directions when leaving region III. The presence of two internal stationary points, of which one is locally stable while the other is locally unstable, means that the isoclines will have two intersections, thus creating the region labeled V in Fig. 5.5B. If the trajectory leaves region III by crossing the algal isocline and entering region IV, it will eventually re-enter region I and trace out the same kind of counterclockwise cyclical orbit as in the previous case. On the other hand, if the trajectory leaves by crossing the grazer isocline and entering region V, grazer biomass will decrease while algal biomass continues to increase, until the system eventually is trapped at the grazer extinction point.

5.4 Extinctions, Periodic Orbits, and Domains of Attraction

The presence of multiple stationary points means that the system can be attracted to different steady states, depending on the initial conditions. The set of initial conditions from which the system will be attracted to a given stationary state is said to be its domain, or basin of attraction.

Extinction or Persistence. If the dilution rate is above the persistence boundary, the system has two locally stable states: the grazer extinction point and the internal equilibrium. When the system is initiated inside a closed region surrounding the internal equilibrium (the shaded area in Fig. 5.6), it will remain inside this basin of attraction and eventually settle down at the internal steady state (Appendix A9 contains details on the computational procedures for locating the boundary of the attraction basin). If the initial state is chosen anywhere outside the domain of attraction for the internal equilibrium, the system will eventually end up at the grazer extinction point. Apparently, ending up at one of these two locally stable equilibria is the only possible fate of any trajectory originating inside the positive cone, when the dilution rate is above the persistence boundary.

The domain of the internal equilibrium has the form of a semilogarithmic ellipse centered at the internal equilibrium. When the input P concentration is increased, the steady-state biomasses at the internal equilibrium will approach the asymptotic levels, C'' and Z', implying that the basin of

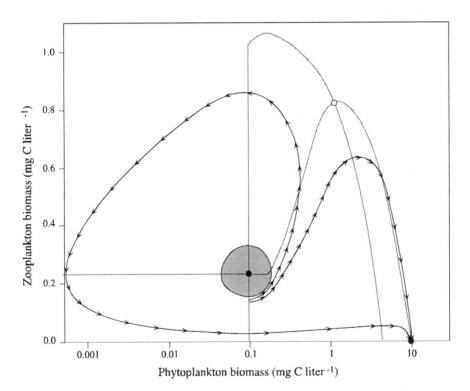

Fig. 5.6. Phase portrait of a nonpersistent system [dilution rate $D = 0.1$ day^{-1}, input P concentration $P_L = 40$ (µg P) l^{-1}]. *Directed paths* are trajectories from two different initial conditions; *shaded area* is the domain of attraction for the internal equilibrium; *broken lines* algal and grazer isoclines; *filled circles* locally stable stationary points; *open circles* locally unstable stationary points

attraction will remain below a fixed size as P loading is increased, which again means that as P input is increased, the attraction basin of the internal equilibrium will cover a diminishing fraction of the positive cone. In other words, there will be an increasing likelihood that a given trajectory, starting from a random initial condition, will lead to extinction of the grazer population when the system is enriched by increasing the P loading.

Focus or Limit Cycle. When the dilution rate is below the persistence boundary, the system has only one locally stable state at the internal equilibrium. Figure 5.7 shows that surrounding this internal equilibrium there will be an elliptic basin of attraction, as in the previous case. Inside the attracting basin, the system will exhibit damped oscillations as it settles down to the internal equilibrium; in other words, the internal equilibrium will be a spiral focus. In contrast to the nonpersistent system, trajectories originating outside the attraction basin of the internal equilibrium will now

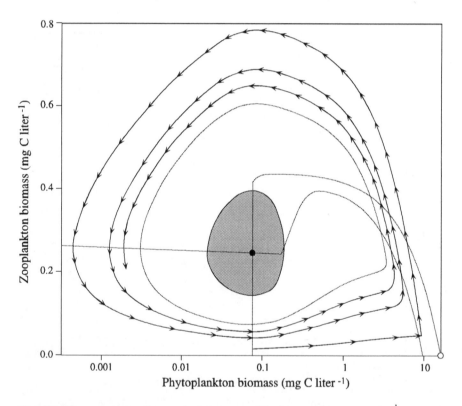

Fig. 5.7. Phase portrait of a persistent system (dilution rate $D = 0.01$ day^{-1}, input P concentration $P_L = 40$ (µg P) l^{-1}]. *Directed paths* are trajectories from two different initial conditions; *shaded area* is the domain of attraction for the internal equilibrium; *broken lines* algal and grazer isoclines; *filled circles* locally stable stationary points; *open circles* locally unstable stationary points

be trapped in a stable periodic orbit, or limit cycle. This means that the system will persist for all initial conditions inside the positive cone, but the stationary state can either be a fixed point or a periodic orbit, depending on the initial conditions. By the same kind of reasoning as in the previous case, we can expect the internal attraction basin to remain below a fixed size as P loading is increased. Thus we should expect the probability for any trajectory, starting from random initial conditions to end up at the internal equilibrium, to diminish with increasing P supply.

Geometry of the Limit Cycle. Sudden, qualitative changes in dynamic behavior as result of changes in model parameters are usually called bifurcation phenomena in the terminology of nonlinear dynamic systems. Many textbooks in theoretical ecology (e.g., May 1975; Roughgarden 1979) emphasize a mechanism called the *Hopf bifurcation* when discussing transitions from stable equilibria to limit cycling behavior in prey-predator models. The Hopf bifurcation results when a change in one or more model parameters causes a local stability change at an equilibrium point (a change from negative to positive real part in a pair of complex eigenvalues). Therefore, the appearance of a limit cycle through a Hopf bifurcation implies the disappearance of a stable equilibrium, and excludes the possibility of coexistence between the two. In the present model, a stable equilibrium is found to coexist with a stable periodic orbit (Fig. 5.7); furthermore, the local stability analysis in Appendix A6 shows that if the internal equilibrium exists, then it will be unconditionally locally stable. In other words, the limit cycle in the present model cannot be generated by a Hopf bifurcation.

Still, the Hopf bifurcation is not the only mechanism by which periodic orbits can be generated in nonlinear dynamic systems (see, for example, Thompson and Stewart 1986 for a general review). If the system possesses two or more unstable stationary points that are saddle points (that is, they have eigenvalues with opposite signs), the system can have closed periodic orbits that are *saddle cycles*. Saddle points have the property that they will be attracting to state trajectories approaching from some directions, while trajectories approaching from other directions will be repelled. The set of points formed by trajectories attracted to or repelled from a stationary point is called the *outstructure* of the point. The subset of the outstructure where trajectories are attracted is called the inset of the stationary point, while the subset where trajectories are repelled is called the outset of the point. It can be shown (e.g., Thompson and Stewart 1986) that eigenvectors corresponding to eigenvalues with negative real parts are tangent to the inset at the stationary point, while eigenvectors corresponding to eigenvalues with positive real parts are tangent to the outset. For an asymptotically stable stationary point, the outstructure will consist only of an inset, which will be identical to the basin of attraction.

When the P loading conditions in the present model are such that both the dilution rate is below the persistence boundary and input concentration is above the threshold for positive zooplankton biomass, both the stationary points at the C axis (the washout point and the grazer extinction point) will be saddle points in the sense that they have two negative and one positive eigenvalues (cf. Appendix A6). In addition to the two unstable saddle points, we will also have a stable equilibrium at the internal focus. From Fig. 5.7, we can see that trajectories approaching the unstable washout point at the origin will be repelled along the C axis, while trajectories

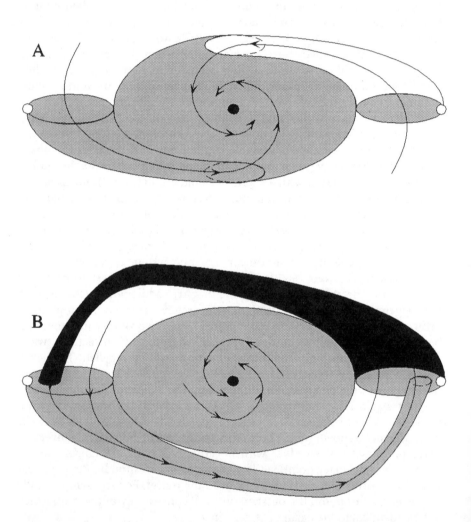

Fig. 5.8A,B. Outstructure geometry of the stationary points at **A** low P loading, and **B** high P loading. *Directed paths* State trajectories; *filled circles* locally stable stationary points; *open circles* locally unstable stationary points

attracted to the unstable grazer extinction point will be repelled in a direction with increasing Z and decreasing C. We can envision two possible fates of trajectories repelled from the saddle points; they can either encounter the attraction basin of the stable focus, or they can get caught in the inset of the opposite saddle point.

We can visualize geometrically the interactions among the outstructures of these three stationary points by representing the outstructures of focal points as ellipsoid surfaces and the outstructures of saddle points as funnel-shaped surfaces (Fig. 5.8). One end of the funnel will correspond to the inset, such that trajectories entering here will first be attracted to the saddle point, and then repelled through the other end of the funnel.

When P loading is low (Fig. 5.8A), the attraction basin of the central focus intercepts the outsets of the saddle points, so that any trajectory repelled from a saddle point is trapped in the domain of the focus. When P loading is increased, the extent of the state space will grow faster than the attraction basin, meaning that more and more of the state space will be taken up by the outstructures of the saddle points. Above a critical P loading, the focal attraction basin no longer intercepts all trajectories in the outsets of the saddle points (Fig. 5.8B), meaning that some trajectories in the outset of one saddle point can be caught in the inset of the opposite saddle point. In other words, the emergence of a limit cycle with increasing P loading in the present model is caused by the connection between the insets and outsets of the two unstable saddle points.

Sensitivity to Initial Conditions. Nonlinear systems have the generic property that system output can be highly sensitive to the initial conditions. In a plankton community of a temperate lake, the initial conditions will typically be the founding populations present at spring overturn. Depending on the species, the founding population can either be overwintering in the water column, or emerge from sediment-dwelling cysts or resting eggs.

Figure 5.9 shows an example of how different a pair of trajectories can be when the initial conditions are located on opposite sides of the boundary of the focal attraction basin (even if the initial conditions differ by only 0.25%). The two simulations follow each other very closely for the first month or so, when suddenly one of them breaks off into the limit cycle, while the other one spirals into the focus. The transition toward the limit cycle is seen to be accompanied by a quadrupling of the cycle period, and an increasing asymmetry in the phytoplankton oscillations with respect to the asymptotic steady-state value. Figure 5.9 illustrates how qualitatively different seasonal trajectories might result, even in the same lake, depending on the recruitment to the founding populations present at the beginning of the growing season. It also offers an explanation for the observation that *Daphnia* populations can alternate between cyclical and stable dynamics in different years in the same lake (McCauley and Murdoch 1987).

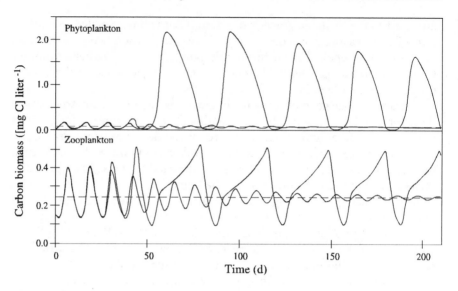

Fig. 5.9. Focus and limit cycle in the time domain. Time courses of phytoplankton and zooplankton biomasses in two different simulations with slightly different initial conditions, both with the same loading conditions $[P_L = 35$ (µg P) l^{-1} and $D = 0.01$ day$^{-1}]$. *Broken horizontal lines* are the corresponding equilibrium biomasses at the internal focus

Cycle Periods, Seasonal Dynamics, and the Spring Clear-Water Phase. The phase portrait in Fig. 5.7 hides the fact that trajectories are not traced out at constant velocity. The overgrazed phase of the limit cycle (with low algal and high grazer biomass) is always substantially shorter than the undergrazed phase (with severe nutrient limitation and low grazing). At the emergence of the limit cycle (when loading exceeds the bifurcation level), the cycle period is close to 30 days, as in Fig. 5.9. When the loading rate is increased further beyond the bifurcation level, the cycle period increases steeply at the same time as the cycle becomes more and more asymmetric (in the sense that the undergrazed phase occupies an increasing fraction of it). Under even modestly eutrophic conditions [P loadings of 0.4 - 0.5 (µg P) l^{-1} day^{-1}], the undergrazed phase can be so long that a single cycle can cover most of the growing season in temperate lakes (5-6 months).

This phenomenon is illustrated in Fig. 5.10 showing temporal algal biomass development in two simulation runs under different loading conditions - one oligotrophic [0.1 (µg P) l^{-1} day^{-1}], and the other eutrophic [0.5 (µg P) l^{-1} day^{-1}]. Initial conditions for both runs in Fig. 5.10 were chosen to resemble an early spring situation with nutrient-saturated phytoplankton growth and low grazing pressure (algal growth rate at 90% of maximal and zooplankton biomass equivalent to 10% of total P). Both simulation runs produce something resembling a spring bloom, with peak algal biomass being ~ fivefold higher in the eutrophic situation. In the oligotrophic run

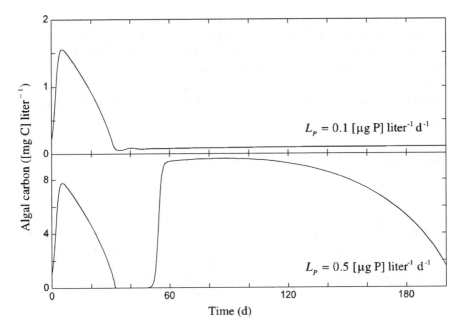

Fig. 5.10. Time courses of phytoplankton biomass in two different simulations with different loading conditions [*upper graph*: $P_L = 10$ (μg P) l^{-1} and $D = 0.01$ day^{-1}; *lower graph*: $P_L = 50$ (μg P) l^{-1} and $D = 0.01$ day^{-1}]. Initial conditions chosen to approximate an early spring situation with low grazing and nutrient-saturated algal growth (algal growth rate at 90% of maximum, grazer biomass corresponding to 10% of total P)

the spring bloom is succeeded by attraction to the stable focus, with grazer control of algal biomass persisting throughout the growing season (set to 200 days in these runs). In contrast, grazer control is maintained for only a short period in the eutrophic run, leading up to a new bloom that endures the whole season.

The latter scenario strongly resembles the so-called *spring clear-water phase* that has been described in several field studies (e.g., Lampert et al. 1986; Vanni and Temte 1990). Different explanations have been offered for the spring clear-water phase, including changes in zooplankton biomass and community structure due to emergence of invertebrate predators or 0+ fish fry (Sommer et al. 1986) and changes in phytoplankton edibility following appearance of grazing-resistant species (Vanni and Temte 1990). The present model shows that it is possible to represent the same pattern without having to take any of these mechanisms into account. The main feature of the present model which produces the prolonged summer-autumn bloom after the spring clear-water phase is the dependency of grazer growth on the nutrient status of algal food; when the algal population recovers after the overgrazing phase, it is able to attain a biomass close to the carrying capacity before the grazers are restored to a biomass level where they can exert an appreciable grazing pressure. When algal biomass is close to the carrying capacity, it will be so nutrient-poor that it will constitute a very low-quality food source for the grazers. Grazers will thus maintain a low growth rate despite the abundance of algal food carbon, leading to a very slow buildup of grazer biomass before it can again constitute a significant loss factor to the phytoplankton.

5.5 Bifurcations and Long-Term Averages

The analysis in the previous section indicates that when the phosphorus loading is sufficiently high, a trajectory originating from a random initial condition will be attracted to a stable periodic orbit with almost unit probability. The bifurcation from a stable focus to a limit cycle can be expected to have strong effects on the long-term averages of the state variables. In Appendix A9 it is shown that the cycle averages of the limit cycle can be computed by a simple augmentation of the state variables. Figure 5.11 shows the computed average biomasses as function of the input P concentration (P_L).

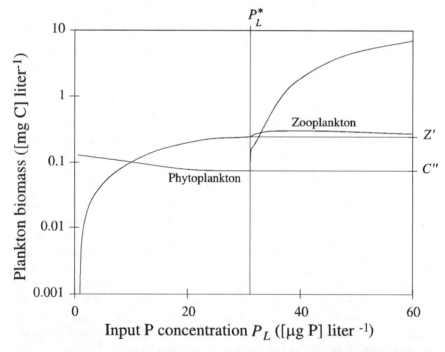

Fig. 5.11. Long-term averages of phytoplankton and zooplankton biomasses as functions of the input P concentration at dilution rate $D = 0.01$ day^{-1}. *Vertical broken line* marks the bifurcation point; *broken horizontal lines* are the analytical solutions for the steady state biomasses at the stable focus

Below the bifurcation level (denoted by P^*_L), computed solutions agree to within numerical precision with the analytical solution at the internal focus. The emergence of the limit cycle leads to only a modest increase in the average zooplankton biomass compared to the asymptotic steady-state value at the internal focus (Z'), reflecting that the zooplankton oscillations are quite symmetrical around the asymptotic steady-state value (Fig. 5.9). On the other hand, traversing the bifurcation level creates a dramatic jump discontinuity in the average phytoplankton biomass, which starts to increase almost linearly with the input P concentration when $P_L > P^*_L$. This can be explained by looking at the time course of the limit cycle (Fig. 5.9), which shows that the maxima and minima of the phytoplankton oscillation are located very asymmetrically with respect to the asymptotic steady-state value, leading to a cycle average that is displaced far above the internal equilibrium.

Carbon Flow Organization. To gain more insight into the dynamics of the limit cycle, we can also compute the cycle averages of different process rates (see Appendix A9). Even though the steady-state algal biomass

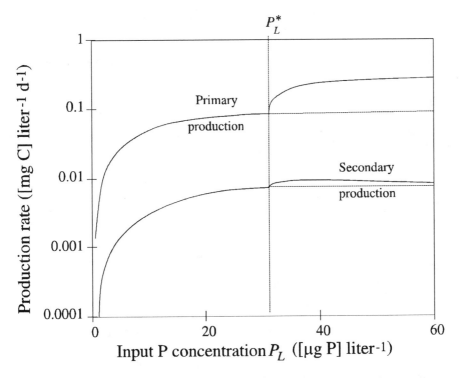

Fig. 5.12. Long-term averages of phytoplankton and zooplankton net production rates as functions of the input P concentration at constant dilution rate $D = 0.01$ day^{-1}. *Vertical broken line* marks the bifurcation point; *broken horizontal lines* are the analytical solutions for the flow rates at the stable focus

decreases with increasing P loading when $P_L < P^*_L$ (cf. Fig. 5.11), this effect on primary production (μC) is masked by the increase in growth rate, such that primary production becomes a monotonously increasing function of P loading (Fig. 5.12). At the internal focus, the equilibrium condition $g = \delta + D$ means that secondary production (gZ) becomes directly proportional to steady-state grazer biomass, which is again proportional to algal growth rate [cf. Eq. (A6.16)].

This common dependency on the algal growth rate explains why both primary and secondary production follow the same pattern in Fig. 5.12, indicating that any increase in primary production caused by an increase in P loading is efficiently channeled into secondary production when $P_L < P^*_L$. Above the bifurcation level, the increase in primary production is not reflected in a corresponding increase in secondary production, indicating that the energy and mass transfer from phytoplankton to zooplankton becomes less efficient. This change in the fate of primary production is perhaps better illustrated in Fig. 5.13, where the total primary production is partitioned into export production (i.e., production lost by sedimentation

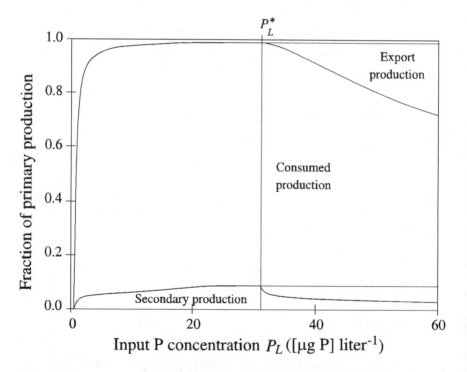

Fig. 5.13. Long-term averages of relative allocation of primary production into export (sedimentation + outflow), consumption and secondary production, as function of the input P concentration at constant dilution rate $D = 0.01$ day^{-1}. *Vertical broken line* marks the bifurcation point; *broken horizontal lines* are the analytical solutions for the flow rates at the stable focus

or outflow) and consumption. The fraction of primary production consumed by zooplankton rises steeply with increasing P loading, approaching an asymptotic level >99%. At the same time, the secondary production per unit of primary production consumed (the trophic efficiency of the grazers) increases as a result of the improving food quality in terms of increasing algal P content. As the input P concentration exceeds the bifurcation level, the trophic efficiency of the grazers drops steeply to approximately half of the asymptotic level (8.6%). At the same time, an increasing fraction of primary production is lost by export processes, instead of being consumed and channeled into secondary production.

Phosphorus Retention. The phosphorus retention (R_p, cf. Section 2.1) is defined as the fraction of P entering the lake that is retained in the lake. From the general expression (2.17) for the P loss rate as function of phytoplankton sinking and zooplankton mortality, the phosphorus retention can be written as

$$R_P = \frac{\sigma P + \delta \theta Z}{D P_L}.$$ (5.8)

The P retention will be constrained between the two limiting cases resulting from setting either $P = 0$ or $Z = 0$ in Eq. (5.8). If phytoplankton sinking is the only loss process, then we find by substituting $Z = 0$ and $P = (1 - R_p) P_L$ into Eq. (5.8), and solving for R_p, that $R_p = \sigma/(D + S)$. At the other extreme, if zooplankton mortality is the only retention process, we find by substituting $P = 0$ and $\theta Z = (1 - R_p) P_L$ into Eq. (5.8), and solving for R_p, that $R_p = \delta/(D + \delta)$.

When the input P concentration P_L is below the critical level P'_L, so that the system is unable to support any grazer population at all, the retention will be $\sigma/(D + \sigma)$, corresponding to phytoplankton sinking alone (Fig. 5.14). Increasing P_L beyond P'_L leads to a rapidly increasing zooplankton fraction of total P, and a corresponding increase in the P retention toward the asymptotic level $\delta/(D + \delta)$. As the zooplankton biomass levels off toward the asymptotic level Z', the zooplankton fraction of total P starts to decrease, resulting in a decreasing P retention with further increase in the P loading. The slight increase in average zooplankton biomass accompanying the transition to the limit cycle at the bifurcation point P^*_L is reflected by an increase in the P retention.

Depending on the partitioning of phosphorus between algae and grazers, the P retention can thus vary by almost an order of magnitude. This result is consistent with the observed increase in P retention following successful biomanipulations (Stenson et al. 1978; Shapiro and Wright 1984; Reinertsen et al. 1989; Sanni and Wærvågen 1990). It also suggests that some of the P retention variability found in lakes with the same flushing rate (Fig. 2.2) might be explained by the phosphorus partitioning in the food web.

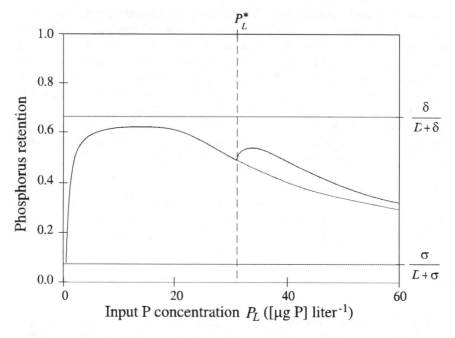

Fig. 5.14. Phosphorus retention as function of the input P concentration at constant dilution rate $D = 0.01$ day^{-1}. *Dashed curve* is the continuation of the stable equilibrium beyond the bifurcation point; *vertical broken line* marks the bifurcation point; *broken horizontal lines* are the P retentions for pure phytoplankton and zooplankton communities

5.6 A Loading Criterion for the Feasibility of Biomanipulation

It appears that for input P concentrations below the bifurcation level, P^*_L, the algal biomass can be said to be strictly grazer-controlled, with any enrichment in terms of increased phosphorus loading being channeled into grazer biomass. As the input level exceeds P^*_L, grazer control is relaxed, with the majority of further phosphorus enrichment remaining in algal biomass, and with an increasing fraction of primary production being lost to the hypolimnion instead of being consumed in the epilimnion. The increased carbon loading to the hypolimnion would be expected to increase the probability of initiating a positive feedback loop leading to accelerated eutrophication through redox-dependent phosphorus release from the sediments. In a qualitative sense, this picture is consistent with the experiences from biomanipulations, as reviewed by Benndorf (1987), suggesting that the bifurcation input P level P^*_L might be an interesting candidate for a practical criterion in evaluating the potential of biomanipulation a given lake.

Sensitivity Analysis. While there appears to be no simple analytical expression for the bifurcation level P^*_L as function of the model parameters, it can still be determined to arbitrary precision by the numerical procedure outlined in Appendix A9. This means that we also can compute numerical approximations to the parameter sensitivities of the bifurcation level for a local region centered at the nominal parameter values (Table 5.1). As the model parameters differ in both units and in magnitude, it is also here reasonable to work with relative parameter sensitivities, or parameter elasticities (i.e., a relative sensitivity of 1 means that a 1% increase in a parameter value increases P^*_L by 1%, while a relative sensitivity of -1 means that a 1% parameter increase gives a 1% decrease in P^*_L).

Figure 5.15 shows that among the algal parameters, it is only the maximum growth rate (μ') that has any effect on the bifurcation level, while the sensitivities to changes in the sinking loss rate (σ) or the subsistence quota (Q') are both negligible. The bifurcation level seems to be less sensitive to grazer parameters controlling growth efficiency [assimilation efficiency (ε) and respiration rate (r)], than to parameters controlling feeding efficiency [incipient limiting food level (C') and maximum ingestion rate (I')], with the mortality rate (δ) in an intermediate position. With the exception of I' and r, all relative parameter sensitivities are positive; for the two most sensitive parameters [incipient limiting food level (C') and grazer P content (θ)], the relative sensitivity is exactly unity, meaning that the bifurcation level is directly proportional to these two parameters. This indicates that the phosphorus economy of the grazer, represented by the parameter θ, can be considered an important factor in determining the upper limit to the loading level where stable grazer control is likely to result.

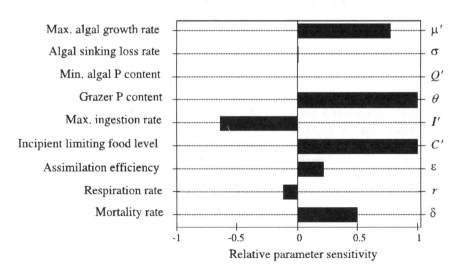

Fig. 5.15. Relative sensitivity of the bifurcation level of input P concentration to changes in different model parameters at dilution rate $D = 0.01$ day^{-1}

Species Substitutions, Tradeoffs, and Robustness. The parameter values in Table 5.1 are deliberately chosen to resemble specific taxonomic groups of phyto- and zooplankton (small, edible, fast-growing algae, like crypto-monads, and large cladoceran filter-feeders with high P demands, like *Daphnia*). The sensitivity analysis in Fig. 5.15 indicates that the bifurcation level P^*_L is quite sensitive to some of the model parameters, so that sub-stitution of parameter values characteristic of other taxonomic or func-tional groups could potentially cause large changes in P^*_L.

In the discussion of interspecific differences among herbivorous zooplankton in Section 4.7, it was argued that there appears to be a tradeoff between high growth rate and low threshold food level for positive popula-tion growth, so that fast-growing species have high threshold levels, and vice versa. If such a tradeoff exists, it should also imply a correlation between the incipient limiting food level (C') and the maximum ingestion rate (I'), which again would imply that the decrease in P^*_L resulting from substituting a grazer with lower C' (like a calanoid copepod instead of a daphnid) would to some extent be counteracted by the corresponding decrease in I'.

The literature survey of interspecific differences among phytoplankton in Section 3.4 indicated a tradeoff between maximum growth capacity and the development of predator defence; species with large colony size, elon-gated cellular form, or other morphological structures that make them more resistant to zooplankton grazing, have significantly lower maximum growth rates than species without such adaptations. If predator defence is limited to morphological adaptations that reduce grazer feeding efficiency, without having any other negative effect on the grazer (excluding possible biochemical defence mechanisms like endotoxins), the main effect on the grazer should be to increase the incipient limiting food level (C') such that a higher algal biomass is needed to saturate the feeding response. This should imply that the decrease in P^*_L resulting from substituting a grazing-resistant alga with lower μ' would to some extent be counteracted by the corresponding increase in C'.

It thus appears that the presence of tradeoffs among competitive traits in candidate phyto- and zooplankton species, would tend to make a loading criterion based on the bifurcation input concentration more robust than if the model parameters were completely independent. This gives hope of developing a loading criterion with some generality beyond the simple two-species target community of the model, consisting of a short, efficient food chain from cryptomonads to daphnids.

The Loading Diagram. Thus far, the model response to changes in the phosphorus loading has been studied by varying the input P concentration P_L, while keeping the dilution rate D at a fixed level. If we define the volu-metric P loading rate as $L_P = D\ P_L\ [(\mu g\ P)\ l^{-1}\ day^{-1}]$, we see that the plankton community can experience the same P supply rate for infinitely

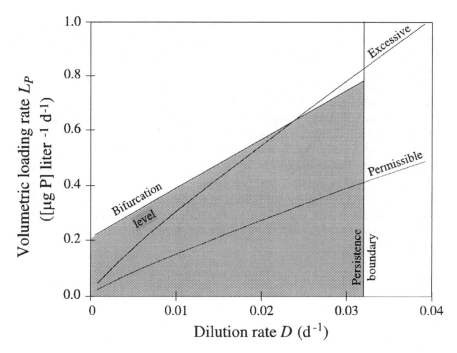

Fig. 5.16. Phosphorus loading diagram: high likelihood of stable grazer control on algal bio-mass within the *shaded region* delimited by the persistence boundary and the bifurcation loading level. Model loading limits are calculated assuming 22% load decay (cf. Fig. 2.2). *Broken lines* marked permissible and excessive are the phosphorus loading limits developed by Vollenweider (1976)

many combinations of input concentration and dilution rate. Since the dilution rate also enters the mass-balance equations of both phyto- and zooplankton as outflow loss terms, systems with the same P supply rate, but with different dilution rate, will not be dynamically equivalent. If we determine the bifurcation level numerically (by the method described in Appendix A9) for different dilution rates up to the persistence boundary (D'), we can display the resulting curve in a classical loading diagram (sensu Vollenweider 1976) with volumetric P loading rate (L_p) on the y-axis and dilution rate (D) on the x-axis.

It turns out that the critical loading rate at the bifurcation point, $L^*_p = D P^*_L$, is very close to a linear function of the dilution rate, and can be represented by a straight line for all practical purposes. This does not mean that P^*_L is a constant, independent of the dilution rate, as the straight line has a positive intercept with the y-axis (Fig. 5.16). In the region above this line (where $L_p > L^*_p$), the probability of ending up in a state with strict grazer control of algal biomass should be low (but still a possible outcome if the initial condition happens to be within the focal attraction basin). In the region below this line (where $L_p < L^*_p$), the system will end up in a stable equilibrium with strict grazer control of algal biomass, unless the dilution rate is above the persistence boundary $(D > D'_c)$; in that case, grazer extinction is most likely to result (except for a small set of initial conditions located within the focal attraction basin).

In order to make the loading diagram compatible with measured loading figures, we will assume that a fraction of the P load is lost before entering the pelagic zone of the lake, as suggested by Fig. 2.2. Using the average load decay of 22%, estimated from the data set used by Prairie (1988), the resulting loading diagram (Fig. 5.16) bears a clear resemblance to the loading limits constructed by Vollenweider (1976). Although the upper limit of the shaded area in Fig. 5.16 intersects the "excessive" loading limit of Vollenweider (1976), this does not necessarily mean that biomanipulation cannot work in eutrophic lakes. In fact, even the most heavily eutrophicated lakes considered by Vollenweider (1976) are within the loading limits for stable grazer control on algal biomass in Fig. 5.16.

Case Studies. As pointed out by Benndorf (1987), the omission of basic limnological information on water renewal and nutrient loading is a major weakness shared by many biomanipulation studies. The set of available data for testing the loading criteria proposed here is therefore very limited. Still, the few documented cases of food web manipulations where reliable loading figures also exist (Fig. 5.17) seem to conform reasonably well with the pattern predicted by the present model in six out of seven cases. However, since the publication of Sanni and Wærvågen (1990), it has been reported that the shallow lake Mosvatn, where the model predicted that biomanipulation should fail, has been invaded by large stands of macrophytes. As such, the biomanipulation of Mosvatn should perhaps be reclassified as unsuccessful, albeit not due to the kind of mechanisms that are considered in the present model.

Benndorf (1987) proposed a biomanipulation-efficiency threshold at an areal phosphorus loading of 0.5-1 (g P) m^{-2} year^{-1} in lakes with mean depths of 5-10 m. This threshold level corresponds quite nicely with the volumetric loading limit around 0.1 (g P) m^{-3} year^{-1} predicted by the present model for lakes with water residence times >1 year. The two lakes considered by Benndorf (1987), Bautzen reservoir and Gräfenhain, are both classified in conformance with the conclusions of Benndorf (1987).

The failure of biomanipulation to improve water quality in the two hypereutrophic Danish lakes, Søbygård Sø and Væng Sø (Jeppesen et al. 1990), is also predicted by the model. The model predicts that the flushing rates in these two lakes are so high that it is unlikely for a persistent grazer population controlling algal biomass to become established, even with major reductions in the external P loading. Of the three lakes in this material that have undergone major reductions in the external P loading (Bautzen Reservoir, Væng Sø, and Gjersjøen), it is only in Gjersjøen that load reduction appears to have improved the odds for successful biomanipulation.

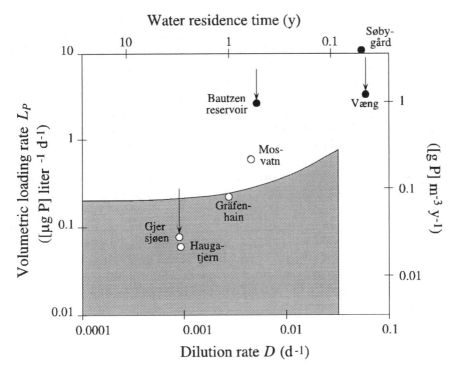

Fig. 5.17. Examples of loading parameters from different lakes in relation to the region of stable grazer control on algal biomass (*shaded area*). *Open symbols* are lakes from which successful biomanipulations have been reported, while *solid symbols* are unsuccessful ones. *Vertical arrows* denote external load reductions resulting from phosphorus abatement measures. (Based on data from Benndorf 1987; Brabrand et al. 1990; Jeppesen et al. 1990; Reinertsen et al. 1989; Sanni and Wærvågen 1990)

5.7 Phosphorus–Biomass Relationships in Lakes

Zooplankton. McCauley and Kalff (1981) showed that in a set of literature data from different lakes there was a significant positive log-log relationship between zooplankton biomass and algal biomass (measured as biovolume). Rognerud and Kjellberg (1984) also found a positive correlation between zooplankton and algal biomass (measured as chlorophyll *a*) in a more homogeneous data set from large, oligo- and mesotrophic Norwegian lakes. The linear equation fitted by Rognerud and Kjellberg (1984) had a steeper slope than the regression model of McCauley and Kalff (1981) and predicted much higher zooplankton levels when extrapolated to the most eutrophic lakes considered by McCauley and Kalff (1981). This inconsistency gives some indication that the relationship between algae and zooplankton might be different in oligotrophic and eutrophic lakes.

The positive correlation between algal and zooplankton biomasses does not necessarily imply that zooplankton biomass is controlled by phytoplankton abundance. It could also be that the relationship is confounded by both variables being controlled by a common third variable like total phosphorus. In a phosphorus-limited lake, available phosphorus could function as a common carrying capacity, setting the upper limit to the potential yields of both phyto- and zooplankton biomasses. Several studies have, in fact, found more of the variance in zooplankton abundance to be explained by total phosphorus, than by algal biomass (Hanson and Peters 1984; Yan 1986; Hessen 1992).

The present model gives predictions on the relationship between phosphorus and zooplankton abundance for the special case where all algal biomass is assumed to be suitable food for zooplankton, and where zooplankton dynamics are assumed to be unaffected by predation from higher trophic levels. As such, the present model should give an approximation of the maximal yield of zooplankton at a given phosphorus supply level. The presence of either inedible algae or zooplankton predators would be expected to reduce the zooplankton biomass below the level predicted by the model.

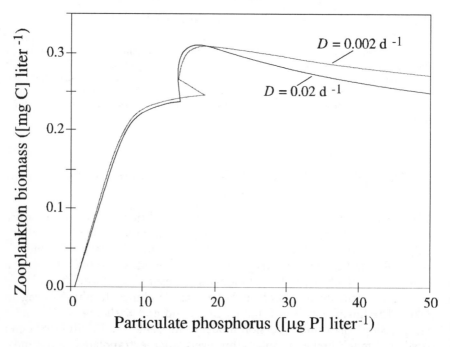

Fig. 5.18. Relationship between average zooplankton biomass and total particulate phosphorus $(P + \theta Z)$ for two different dilution rates (D)

In the preceding sections, model output has been presented as function of input concentration, although some emphasis has also been placed on the interaction between water flow and input concentration in determining the critical P loading level for stable grazer control of algal biomass. This presentation form hides the fact that for different loading levels, leading to the same average level of total particulate P (i.e., $P + \theta Z$), the average phyto- and zooplankton biomasses will be quite similar. Figure 5.18 shows, as an example, the predicted relationships between grazer biomass and total particulate P for two dilution rates differing by an order of magnitude. The difference in the two curves is caused by the increasing importance of algal sinking losses to the phosphorus retention with increasing water residence time. Figure 5.18 indicates that the phosphorus–biomass relationships predicted by the present model are insensitive to differences in the flushing rate (D), and that they therefore should apply across lakes with different hydraulical characteristics.

NIVA made an impressive survey of plankton biomass and water chemistry in 355 Norwegian lakes, ranging from ultraoligotrophic to hypereutrophic in temperate to arctic and subalpine areas (Faafeng et al. 1990). One problem with relating the results from the NIVA study to the model predictions is that only total phosphorus was measured. This means that some decision must be made on how the forms of phosphorus considered by the model relate to measured total phosphorus; or more specifically, whether dissolved organic phosphorus (DOP) can be considered an active part of the pelagic phosphorus cycle, or not. The low variability in the relative DOP fraction of total P (Fig. 2.4), could be taken as an argument for both views. If measured DOP is biased by leakage of cell contents from broken cells (Taylor and Lean 1991), it should certainly be considered as a biologically active form. On the other hand, if the partitioning is the result of a slow exchange process with the biologically active forms of phosphorus, DOP should not be part of the phosphorus cycle represented by the present model.

Figure 5.19 shows the relationship between zooplankton biomass and total particulate P predicted by the present model. The main feature of the model prediction is that the relationship appears to be hyperbolic rather than allometric, as assumed by most regression models. The vertical asymptote of the hyperbolic relationship corresponds to the critical input P concentration P'_I, below which the system is unable to support a second trophic level, while the horizontal asymptote corresponds to the asymptotic zooplankton biomass Z'. When data from the NIVA survey are recalculated under the assumption that the DOP fraction of total P does not participate directly in the pelagic phosphorus cycle, the zooplankton biomass is seen to reach the predicted upper limit in some lakes, while remaining more than two orders of magnitude below the potential yield in other lakes. Making the opposite assumption with regards to availability of DOP would displace the data points to the right in Fig. 5.19, but otherwise preserve this pattern.

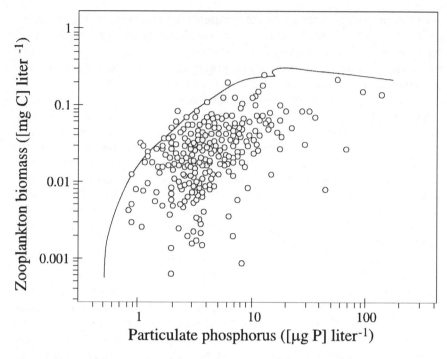

Fig. 5.19. Zooplankton biomass as function of particulate phosphorus (estimated from total phosphorus by assuming 42% DOP). Seasonal averages from 355 Norwegian lakes (Faafeng et al. 1990). *Solid line* is the relationship predicted by the present model

Chlorophyll a. In contrast to particulate carbon, which is constituted by a variable mixture of algae, bacteria, microzooplankton, and detritus, chlorophyll *a* (chla) is specific constituent of living plants. Due to the simplicity, specificity, and sensitivity of chlorophyll *a* measurements, the chlorophyll concentration is therefore often used almost as a synonym for algal biomass. The correlation between chlorophyll *a* concentration and total phosphorus, first reported by Sakamoto (1966), has become a cornerstone of practical lake management. It has also been celebrated as an important example of emergent properties in lake ecosystems (e.g., Harris 1986).

When confronted with results from algal physiology, the implications of a correlation between chla and total P become even more eloquent. Most culture studies on plankton algae show that the ratio between chlorophyll *a* and cell carbon can vary by more than an order of magnitude, depending on the growth conditions (e.g., Geider 1987). In other words, the observed correlation between chla and total P will be reflected by a corresponding relation between algal biomass and total phosphorus only if the growth conditions are similar.

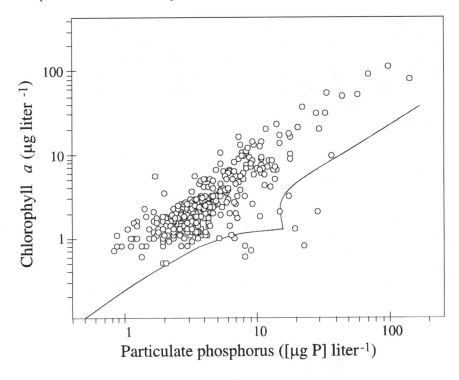

Fig. 5.20. Chlorophyll a concentration as function of particulate phosphorus (estimated from total phosphorus by assuming 42% DOP). Seasonal averages from 355 Norwegian lakes (Faafeng et al. 1990). *Solid line* is the relationship predicted by the present model, using the chla:C relationship given by Eq. (3.16) with $\varphi = 18$ (µg chla) (mg C)$^{-1}$ and $\mu_0 = 0.05$ day^{-1}

If the chlorophyll a concentration is given by φC with φ, the chla:carbon ratio, given by Eq. (3.16) and parameters in Section 3.5, we can obtain the cycle average chlorophyll a concentration by using the method outlined in Appendix A9 for computing averages of functions of the state variables. Figure 5.20 shows that the model predicts an allometric relationship between the long-term averages of chlorophyll a and total particulate phosphorus, with a distinct notch at the transition between stable focus and limit cycle. Although the equilibrium algal biomass is predicted to be a decreasing function of phosphorus supply below the bifurcation point (Fig. 5.11), this effect is overridden by the increase in the chla:C ratio caused by the increase in algal growth rate.

The results from the NIVA survey (making the same assumptions concerning DOP as in Fig. 5.19) show that the majority of the investigated lakes are located above the predicted chlorophyll a–phosphorus relationship (Fig. 5.20). The chlorophyll concentration is seen to reach the predicted lower limit in some lakes, while being more than an order of magnitude higher than the level corresponding to grazer control of algal biomass in other lakes. It is interesting to notice that the notch at the bifurcation point in the predicted chla – P relationship appears to be reflected in the data from the NIVA survey. The location of the notch in terms of total P [20-30 (μg P) l^{-1}, when assuming 42% DOP], is also roughly comparable to the phosphorus level commonly used by limnologists to distinguish between mesotrophic and eutrophic lakes.

5.8 Summary and Conclusions

By this chapter, we have reached the point where we can assemble the submodels developed in the preceding chapters, describing phosphorus cycling and plankton growth dynamics. By radical aggregation of compartments, a minimal model of pelagic phosphorus cycling with only three state variables (algal phosphorus, algal biomass, and grazer biomass) has been constructed. This model is sufficiently simple for many of the model properties to be explored by analytical techniques.

The methods of local stability analysis show that the model can have only one locally stable equilibrium point when the dilution rate is below a certain limit. If the dilution rate exceeds this limit, called the persistence boundary, the system will have two locally stable equilibria, one of which represents the extinction of the grazer population. The possibility of deterministic extinction of the grazer population results from the consideration of stoichiometric relationships between predators and their prey. In this respect, the present model is different from the models considered by Rosenzweig (1971), where deterministic extinction of the predator population is impossible.

Isocline analysis combined with simulations showed that the model has two principal modes of dynamic behavior for dilution rates below the persistence boundary; the system could either be attracted to the central focus predicted by the local stability analysis, or it could end up in a limit cycle. The possibility of a limit cycle behavior is not indicated by the local stability analysis, pointing to the limitations of this technique when the state equations include functions with discontinuous derivatives, like Eqs. (5.5) and (5.6).

The bifurcation sequence leading to the limit cycle corresponds to the paradox of enrichment (Rosenzweig 1971), where a stable equilibrium is destabilized into a limit cycle by increasing the carrying capacity of the prey population. Below a certain phosphorus loading, called the bifurcation level, attraction to the stable equilibrium is the only possible outcome. The emergence of the limit cycle above the bifurcation loading level makes it possible for the system to be attracted to two alternate stationary states, depending on the initial conditions. This coexistence of the two attractors indicates that the limit cycle is not generated by the destabilization of an equilibrium point through a Hopf bifurcation. Instead, it is suggested that the limit cycle results from the interaction between the outstructures of two dynamically unstable saddle points.

The long-term averages of state variables and process rates reveal large differences in phosphorus partitioning, biomass proportions, and organization of carbon flows, between the stable equilibrium and the limit cycle. The emergence of the limit cycle makes the average algal biomass increase steeply with phosphorus loading, while the zooplankton biomass shows only a minor increase above the equilibrium level. This change in biomass proportions is reflected in an increasing fraction of primary production being exported through the outflow or by sedimentation, instead of being consumed and channeled into secondary production. Changes in the partitioning of phosphorus between phyto- and zooplankton create a variation of nearly an order of magnitude in the phosphorus retention of the system, thus demonstrating the potential of zooplankton as a sink for phosphorus.

Due to its dramatic effects on biomass proportions and flow organization, the bifurcation loading level can be proposed as a threshold for the establishment of stable grazer control on algal biomass, which is usually stated as the goal of lake restoration by food web manipulation. It is shown that the bifurcation level and the persistence boundary can be used to construct a loading diagram for the feasibility of biomanipulation, in terms of volumetric phosphorus loading rate and water renewal time of a given lake. For lakes with water renewal times longer than 1 year, the critical loading level is close to 0.1 (g P) m^{-3} $year^{-1}$, thus providing a mechanistic justification for the empirical threshold level proposed by Benndorf (1987). The loading diagram is able to classify correctly the success or failure of six out of seven well-documented biomanipulation experiments.

The biomass–phosphorus relationships predicted by the present model are shown to conform well with observations, in the sense that they provide upper and lower limits to the zooplankton biomass and chlorophyll a concentration at a given phosphorus level. This suggests that the present model might be used to assess the degree of grazer control on algal biomass in a given lake from the average levels of particulate phosphorus, chlorophyll a, and zooplankton biomass.

Even if the minimal model presented here has no representation of seasonality in forcing functions like light and temperature (since all model parameters are assumed time-independent), it is still able to reproduce some characteristic seasonal patterns like the spring bloom and the spring clear-water phase. One can only speculate whether this may indicate that perhaps many of the phenomena we perceive as driven by seasonal forcing are actually created by the interplay between specific initial conditions and simple, non-seasonal dynamics.

5.9 Symbols, Definitions, and Units

Symbol	Definition	Unit
C	Algal carbon biomass	$(\text{mg C}) \, l^{-1}$
C'	Incipient limiting food concentration	$(\text{mg C}) \, l^{-1}$
C''	Threshold food level for positive grazer growth	$(\text{mg C}) \, l^{-1}$
δ	Specific grazer mortality rate	day^{-1}
D	Dilution rate	day^{-1}
D'_-	Dilution rate at the persistence boundary	day^{-1}
ε	Food carbon assimilation efficiency	–
φ	Algal chlorophyll:carbon ratio	$(\mu\text{g chla}) \, (\text{mg C})^{-1}$
g	Grazer growth rate	day^{-1}
g'	Maximum grazer growth rate on optimally composed food	day^{-1}
I	Specific food ingestion rate	day^{-1}
I'	Maximum specific ingestion rate at saturating food	day^{-1}
L_P	Volumetric P loading rate	$(\mu\text{g P}) \, l^{-1} \, \text{day}^{-1}$
L_P^*	Volumetric P loading rate at the bifurcation point	$(\mu\text{g P}) \, l^{-1} \, \text{day}^{-1}$
μ	Algal growth rate	day^{-1}
μ'	Asymptotic algal growth rate	day^{-1}
θ	Grazer P content	$(\mu\text{g P}) \, (\text{mg C})^{-1}$
Q	Algal P content	$(\mu\text{g P}) \, (\text{mg C})^{-1}$
Q'	Minimum algal P content (subsistence quota)	$(\mu\text{g P}) \, (\text{mg C})^{-1}$
P	Concentration of algal-bound P	$(\mu\text{g P}) \, l^{-1}$
P_L	Input P concentration	$(\mu\text{g P}) \, l^{-1}$
P'_L	Minimum input P concentration for $Z > 0$	$(\mu\text{g P}) \, l^{-1}$
P_L^*	Input P concentration at the bifurcation point	$(\mu\text{g P}) \, l^{-1}$
ρ	Specific P release rate	$(\mu\text{g P}) \, (\text{mg C})^{-1} \, \text{day}^{-1}$
r	Specific respiration rate	day^{-1}
R_P	Phosphorus retention	–
σ	Algal sinking loss rate	day^{-1}
Z	Grazer carbon biomass	$(\text{mg C}) \, l^{-1}$
Z'	Asymptotic grazer carbon biomass	$(\text{mg C}) \, l^{-1}$

6 Approaching Planktonic Food Webs: Competition, Coexistence, and Chaos

...both theoretical and experimental analyses have produced a number of mechanisms that permit coexistence despite competition. Therefore, the time has come to turn Hutchinson's paradox upside down and to ask the reverse question, "Why are there so few species of phytoplankton in a given lake?" ... the large discrepancy between the actual number of phytoplankton species in a lake and the maximal possible one shows that it is at least a plausible hypothesis, that quite a lot of competitive exclusion has occurred already.

Ulrich Sommer (1989b).

Resource competition is defined as an indirect form of competition where the participating species are exploiting a shared resource. An increase in one species will, under this condition, decrease the amount of resource available to the other species, and thereby reduce their chances to proliferate. If the resource availability is reduced to such a low level that a species is unable to maintain non-negative net growth rate, this species will eventually be competitively excluded from the community. For an arbitrary number of species competing for a single resource in a constant environment, it can be shown that all but one species will be competitively excluded. This result can be stated more generally as *the principle of competitive exclusion*; that two species can coexist only if they exploit their environment differently (Hardin 1960), or that an arbitrary number of species can coexist at constant levels, if they are limited by the same number of resources (Levin 1970).

The competitive exclusion principle lead Hutchinson (1961) to formulate *the paradox of the plankton*: how can so many species of phytoplankton apparently coexist in a homogeneous epilimnion on so few resources (i.e., light, temperature, and a few mineral nutrients)? Since then, explaining the plankton paradox has become one of the favorite activities among theoretically inclined ecologists. Most explanations that have been offered point out that the simple n species–n resources relationship has been proven only for species coexisting at constant densities in a constant environment, and that either environmental variability or internally generated density fluctuations could promote the coexistence of more species.

In a variable environment, the resource fluctuations themselves become a resource, allowing two or more species to coexist on a single resource (Levins 1979). This theory has been tested for phytoplankton communities by Sommer (1984), who found that a higher species diversity could be maintained in cultures receiving a variable (pulsed) nutrient supply than in those receiving a continuous supply. The idea of coexistence mediated by temporal variability has been extended by Tilman (1982), who showed how an arbitrary number of species might coexist on two fluctuating essential resources, if the species exhibited a tradeoff in their competitive ability for the two resources (that is, no species is the superior competitor for both resources).

In a simple Lotka-Volterra model with two prey and one predator species, Vance (1978) found that predator-mediated coexistence of two prey species could result if the competitors were sufficiently different in at least one of several aspects, such as spatial refuge from the predator, predator selectivity, or allocation between competitive and predatory defense. Armstrong and McGehee (1980) showed in more general terms how several species could coexist on a smaller number of resources when the system has an attractor that is more complex than a simple equilibrium point, supporting persistent fluctuations in prey and predator densities. Gilpin (1979) showed that for certain parameter ranges, even the simple three-species model of Vance (1978) exhibits coexistence with very complex dynamical behavior in a characteristic period-doubling cascade from a stable periodic orbit to spiral chaos.

It should be noticed that these two lines of reasoning differ in the sense that they focus either on fluctuations in resource supply rates due to *externally* forced environmental variability, or on fluctuations in competitor loss rates due to *internally* generated predator variability. In their interaction with the species at the base of the food web, grazers function both as a resource sink by selectively removing prey, and as a source by supplying recycled resource. Thus the activity of grazers could have a dual effect on competition and coexistence among their prey species by creating fluctuations both in prey loss rates and in resource supply rates.

6.1 Eutrophication as an r–K Selection Gradient

In Section 3.3 it was shown that several different models for nutrient uptake and nutrient-limited phytoplankton growth are all equivalent to the Monod model in the steady-state situation. Under the simplifying assumption of no efflux of nutrients ($S'= 0$), the Monod relationship [Eq. (3.10)] between specific growth rate (μ) and the equilibrium concentration of dissolved nutrient (S), resulting when nutrient uptake and growth are in balance, can be written as

$$\mu = \mu'' \frac{S}{K' + S},\tag{6.1}$$

where μ'' is the maximal growth rate and K' is the Monod half-saturation parameter ($S = K'$ is equivalent to $\mu = \frac{1}{2}\,\mu''$).

For nutrient-limited plankton algae, the characteristics of *r*- and *K*-selection sensu MacArthur and Wilson (1967) are usually identified with the parameters of the Monod equation, with *r*-strategists having high maximal growth rates (μ'') and *K*-strategists having low Monod half-saturation parameters (K'). A tradeoff between high μ'' and low K' would imply that the Monod curves of *r*- and *K*-strategists can cross each other. Intersecting Monod curves should theoretically allow the partitioning of species along a resource supply gradient, as first pointed out by Dugdale (1967).

Although resource partitioning in gradients of supply or loss rates appears to be a common phenomenon for bacteria (see Turpin 1988, and references therein), the case for phytoplankton algae remain somewhat ambiguous. Dunstan and Tenore (1974), Harrison and Davis (1979), Mickelson et al. (1979), and Sommer (1986) found species replacements, which could be indicative of a *r-K* tradeoff, when culturing natural phytoplankton communities in a gradient of dilution rates. On the other hand, Smith and Kalff (1983) found that the same species (*Synedra acus*) dominated at all dilution rates up to 0.9 day^{-1}, although the interpretation of their results has been subject to some debate (Sommer and Kilham 1985; Smith and Kalff 1985). The literature data compiled in Chapter 3 shows no clearcut evidence of a positive correlation between the Monod parameters that would be indicative of an *r-K* tradeoff, although this is perhaps not so unexpected considering the wide range of experimental methods and growth conditions employed in different studies.

Model Equations. In order to investigate the effects of grazing on the competitive ability of the three model species introduced in Chapter 3 (Table 3.2), we can apply the general equations describing pelagic phosphorus cycling developed in Section 2.5. If we consider the case of only one nonselective grazer population, the Eqs. (2.7) and (2.8), describing the dynamics the carbon biomass of species i [C_i; (mg C) l^{-1}] and the concentration of phosphorus contained in species i [P_i; (μg P) l^{-1}], can be simplified to

$$\dot{C}_i = \left(\mu_i - (D + \sigma + FZ)\right)C_i \tag{6.2}$$

$$\dot{P}_i = v_i C_i - (D + \sigma + FZ)P_i \tag{6.3}$$

for $i = 1, 2, 3$. The specific growth rate of species i (μ_i; day^{-1}) is described by the Droop model [Eq. (3.2)], with the cellular phosphorus quota of species i given by $Q_i = P_i/C_i$, and with parameters as in Table 3.2. The specific P uptake rate of species i [v_i; (μg P) (mg C)$^{-1}$ day^{-1}] is described by first-order uptake kinetics as in Eq. (3.4), with the uptake affinity a linear decreasing function of the P quota [Eq. (3.5)], and with the parameters given in Table 3.2.

Under the assumption of non-selective grazing and equal sinking loss rates (σ; day^{-1}), all species will suffer the same loss rate at a given grazer biomass level [Z; (mg C) l^{-1}]. If we define the total amount of food carbon available for the grazers as $C = C_1 + C_2 + C_3$, the specific clearance rate [F; l (mg C)$^{-1}$ day^{-1}] will be $F = I/C$, where I (day^{-1}) is the specific ingestion rate of the grazers. The specific ingestion rate can be assumed to be described by the same piecewise linear function of total available food concentration as in Eq. (5.5), with the same parameters as in Table 5.1. The dynamics of the grazer population will be determined by the balance between growth and losses due to mortality (δ; day^{-1}) and dilution (D; day^{-1}), as in Eq. (5.1):

$$\dot{Z} = \left(g - \left(D + \delta\right)\right)Z .\qquad (6.4)$$

By defining the food phosphorus available for the grazers as $P = P_1 + P_2 + P_3$, the equation describing grazer growth rate (g; day^{-1}) as function of food P and C concentrations becomes identical to Eq. (5.6), with the same parameters as in Table 5.1. If we denote the dissolved inorganic P concentration by S [(μg P) l^{-1}], the mass-balance equation for dissolved inorganic P [Eq. (2.13)] becomes

$$\dot{S} = D\left(P_L - S\right) - \sum_i v_i C_i + \rho Z .\qquad (6.5)$$

Nonselective grazing implies that the specific P recycling rate of the grazers [ρ; (μg P) (mg C)$^{-1}$ day^{-1}), given by Eq. (2.14), can be simplified to $\rho = F P - g\theta$, where θ is the P content of the grazers [(μg P) (mg C)$^{-1}$].

Species Replacements in a Phosphorus Loading Gradient. In order to have coexistence at a fixed equilibrium point with all three phytoplankton species having nonzero biomasses, we must require that all derivatives in the mass-balance equations (6.2)-(6.5) vanish. Since the assumption of nonselective grazing implies that the loss terms for all three algal species must be equal, we must also require that all the equilibrium growth rates are equal ($\mu_1 = \mu_2 = \mu_3 = D + \sigma + F Z$). Substituting $D + \sigma + F Z = \mu_i$ into Eq. (6.3) gives $v_i C_i = \mu_i P_i$ at the equilibrium point. Rearranging this to $v_i = \mu_i Q_i$ shows that all species must satisfy the condition for balanced growth [Eq. (3.6)] at the equilibrium point, and that all species must therefore satisfy their respective Monod equations. This means that we will have three different Monod equations relating the equilibrium value of S to the same steady-state growth rate, and that these three equations cannot be consistent unless all three species have identical Monod parameters. This proves that we cannot have stable coexistence of the three model species in the presence of a nonselective grazer, and that only one algal species can have non-zero biomass at a given equilibrium point of the equation system (6.2)-(6.5).

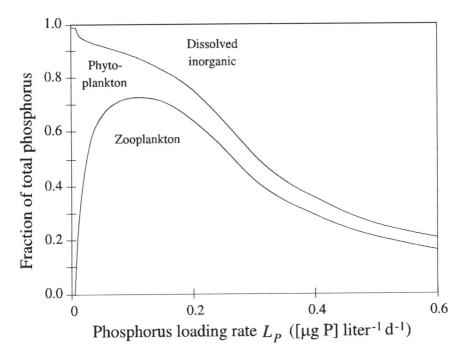

Fig. 6.1. Fraction of total phosphorus in grazers, algae (only species 2), and dissolved inorganic P at the stable focus of the equation system (6.2)-(6.5), as function of phosphorus loading (at a constant dilution rate $D = 0.01$ day^{-1})

The main features of the two-species equilibrium point, or stable focus, of system (6.2) - (6.5) when only one phytoplankton species is given non-zero initial conditions, are shown in Fig. 6.1. At low phosphorus loading $[L_P = DP_L;$ (μg P) l^{-1} day^{-1}] the system will end up in a stable equilibrium point, or focus, from an arbitrary initial condition. If the P loading then is slowly increased, the system will remain at the stable focus, indicating that the focus has a nonzero attraction basin over the whole loading range, as in Fig. 5.7. Above a critical lower loading limit, the system becomes sufficiently productive to support a second trophic level. With further enrichment, the zooplankton fraction of total phosphorus increases steeply, while the algal fraction decreases correspondingly. The inorganic fraction of total P also increases, reflecting the increase in algal equilibrium growth rate caused by the enrichment. As algal growth rate approaches the maximum (μ''), the equilibrium phyto- and zooplankton biomasses approach asymptotic levels, as in Fig. 5.1. With the algal P quota being constrained below Q'' in the present model, an increasing fraction of total P will be allocated to inorganic P as μ approaches μ'' and the biomass levels tend to their asymptotes.

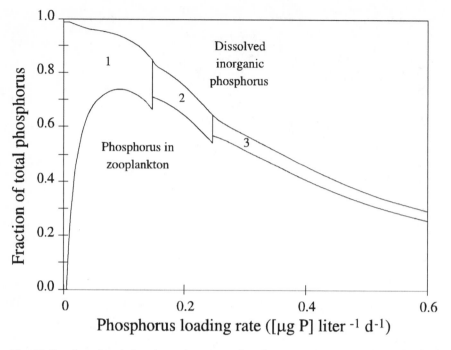

Fig. 6.2. Fraction of total phosphorus in grazers, algae (species 1, 2, and 3), and dissolved inorganic P at the stable focus of the system described by Eqs. (6.2)-(6.5), as function of phosphorus loading (at a constant dilution rate $D = 0.01$ day^{-1})

If we consider the eutrophication gradient as a gradient in steady-state phytoplankton growth rates, we would expect a replacement sequence of species in an r–K continuum with increasing phosphorus loading. If we start at the stable equilibrium point with only species 2 resident (as in Fig. 6.1) and introduce small inocula of species 1 and 3 (1% of species 2 biomass), we can perform a sequence of simulated invasion experiments. Figure 6.2 shows that species 1 can invade this equilibrium at low loading rates and species 3 at high loading rates, while species 2 remains resident at intermediate levels (the same figure would result if we chose species 1 or 3 as the resident). Figure 6.2 also shows that the species transitions are abrupt without any intervening interval of coexistence. The species replacements are accompanied by jump discontinuities in the zooplankton fraction of total phosphorus, resulting from the faster-growing invading species being able to support a higher asymptotic zooplankton biomass level. Otherwise, the partitioning of total P between dissolved inorganic, algae, and grazers is very similar to Fig. 6.1.

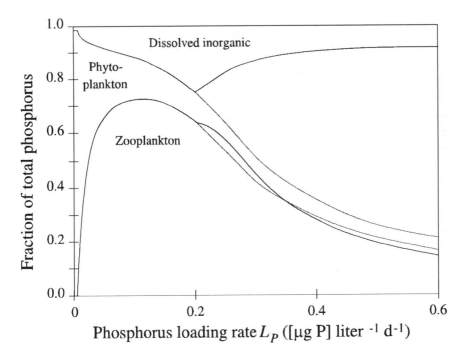

Fig. 6.3. Average fraction of total phosphorus in grazers, algae (only species 2), and dissolved inorganic P in the limit cycle of the equation system (6.2)-(6.5) as function of phosphorus loading (at a constant dilution rate $D = 0.01$ day^{-1}). *Dashed lines* represent the equilibrium solution from Fig. 6.2

Like the model analyzed in Chapter 5, the present model also possesses an alternate stationary state in the form of a limit cycle. When starting from a random initial condition at a high phosphorus loading rate, there is a high probability that the system will be attracted to the limit cycle and not the stable focus. If the loading rate then is slowly decreased, the system will remain in the limit cycle until some critical loading level, called the bifurcation level in Chapter 5, where the limit cycle converges to a stable equilibrium point. The resulting long-term average phosphorus partitioning for a system with only species 2 present is shown in Fig. 6.3. The main difference between the two modes of dynamic behavior is that the limit cycle allows a much higher utilization of dissolved inorganic P by the phytoplankton, while the fraction of total P allocated to zooplankton is practically the same. The reduced average inorganic P concentration implies that the average algal growth rate is lower in the limit cycle than at the stable equilibrium, with a maximum at the bifurcation loading level.

Repeating the invasion experiments from Fig. 6.2 with species 2 resident in a limit cycle (Fig. 6.4), results in a qualitatively different pattern for loading rates above the bifurcation level. Below the bifurcation level, the invasion of

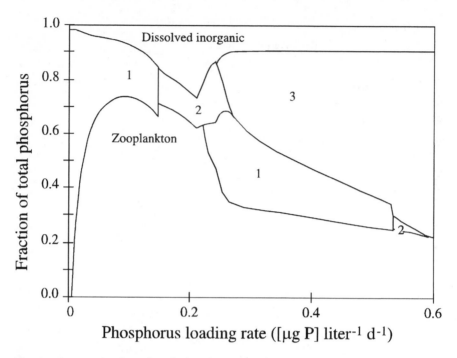

Fig. 6.4. Average fraction of total phosphorus in grazers, algae (species 1, 2, and 3), and dissolved inorganic P in the limit cycle of the system (6.2)-(6.5), as function of phosphorus loading (at a constant dilution rate $D = 0.01$ day^{-1})

species 1 at low loading rates will be the same as in Fig. 6.2. With the emergence of the limit cycle, both species 1 and 3 can successfully invade and become coexistent with the resident species. With further increase in the P loading, the three-species coexistence is disrupted by the exclusion of species 2, while the two extremes in the r–K gradient, species 1 and 3, continue to coexist. Species 3 more and more displaces species 1 with further enrichment, until suddenly species 1 is replaced by species 2, which then coexists as subdominant to species 3 over a small loading interval.

With increasing phosphorus enrichment, the limit cycle oscillations become increasingly more violent, with fast shifts from an overgrazed situation with high turnover of algal biomass and high nutrient availability to an undergrazed situation with slow algal growth and severe nutrient limitation. The relative duration of the intermediate situations, which should be most favorable to species 2, becomes progressively shorter, leading to the exclusion of species 2 from the three-species coexistence.

The phase portrait of the limit cycle in Fig. 6.5 shows how the growth phases of species 1 and 3 exploit different parts of the limit cycle. When zooplankton biomass declines to a level where it can no longer control algal biomass, the r-strategist, species 3, is the first to exploit the situation. As

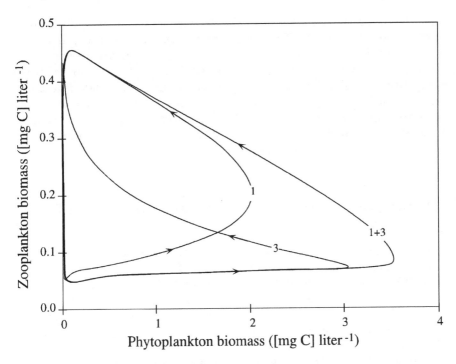

Fig. 6.5. Phase portrait of phyto- and zooplankton biomasses in the limit cycle of the system (6.2)-(6.5) at loading rate $Lp = 0.3$ (μg P) l^{-1} day^{-1} and dilution rate $D = 0.01$ day^{-1}, showing orbits for species 1 and 3, and their sum (1 + 3)

the species 3 population increases, it depletes the initially available inorganic P pool, and becomes progressively more phosphorus-limited. With increasing P-limitation of species 3, the competitive advantage of the *K*-strategist, species 1, also increases, leading to the gradual displacement of species 3. In the meantime, zooplankton biomass has been accumulating to a level where none of the species can maintain positive net growth, thus taking the system back to the overgrazed situation.

The increase in zooplankton biomass with increasing phosphorus loading makes the eutrophication gradient equivalent to a gradient in equilibrium grazing loss rates. When the system is attracted to the internal focus, only equilibria with a single prey species are possible, creating a sequence of phytoplankton species replacements with increasing nutrient enrichment. In contrast, the emergence of a limit cycle allows the coexistence of up to three phytoplankton species on a single limiting resource. The alternations between unlimited growth and extreme nutrient limitation in the periodic orbit of Fig. 6.5 can be viewed as two temporally separated niches that can be exploited by either the *r*- or the *K*-strategist in a pair of competitors. The resulting coexistence of two species on a single resource therefore does not violate the competitive exclusion principle if we use the formulation of Hardin (1960; two species can coexist if they exploit the environment differently). This interpretation also explains why it is the extremes of the r-K continuum (species 1 and 3) that form a stable coexistence in the domain of the limit cycle, while species with intermediate competitive abilities (like species 2) are competitively excluded from both temporal niches (except for a limited range of loading rates). The exclusion of intermediate species makes it unlikely that the dynamics of the simple limit cycle can support the coexistence of more than a few phytoplankton species from a unidimensional r-K continuum.

6.2 Differential Loss Rates and Invadability of Equilibria

Phytoplankton ecologists have traditionally been divided into two different schools, emphasizing either differential reproductive rates (e.g., Tilman et al. 1982) or differential loss rates (e.g., Kalff and Knoechel 1978) as the dominating forces behind phytoplankton succession. Still, a balanced view seems to be emerging, where it is recognized that differential loss processes can play an important role by tilting the balance of resource competition (Kilham 1987; Sommer 1988b).

Differential losses result when competing phytoplankton species have different susceptibilities to grazing, or different abilities to remain suspended in the pelagic zone. These two major loss processes differ fundamentally in that the impact of selective grazing can be expected to be density-dependent, while differential sinking losses will generally be density-independent. Furthermore, while sinking is always a net loss of nutrients from the pelagic zone, the nutrient content of grazed phytoplankton cells will either be immediately recycled or remain suspended for some time as a constituent of grazer biomass.

Many phytoplankton species show morphological or biochemical adaptations that can be interpreted as forms of predator defence. Increasing size is generally accepted as an advantageous adaptation against consumption by common filter feeders, but because it also reduces the relative adsorptive surface area as well as increases the sinking rate (Reynolds 1984), one would expect this strategy of predator defence to have its costs in terms of reduced competitive ability for essential resources and increased sinking losses. It thus appears that no single cell shape can be optimal for all combinations of loss and nutrient supply rates, and that phytoplankton species seem to be facing a tradeoff between minimizing loss rates or maximizing growth rates in order to maximize their evolutionary fitness (cf. Kilham and Hecky 1988).

Water quality improvement from food chain manipulations depends on zooplankton-induced mortality rates being sufficient to match the growth rates of all species in the phytoplankton community (Gliwicz 1990). Invasion or persistence of grazing-resistant phytoplankton species has been identified with the failure of several biomanipulation experiments to produce the anticipated reduction in algal biomass and increase in water transparency (Lynch 1980b; Benndorf et al. 1984; Jeppesen et. al. 1990). Since the group of grazing-resistant algae contains several nuisance-bloom species like colonial cyanobacteria, one can easily imagine situations where a failed biomanipulation actually would lead to further deterioration of water quality. One would therefore expect the long-term stability of biomanipulated lake communities to be crucially dependent on the persistence of a short, efficient food chain structure with high resistance to invasion by inedible phytoplankton species.

The Mathematics of Selective Feeding. According to the definition of Jacobs (1974), selective feeding is said to take place when a predator consumes cooccurring prey species at different rates. In the absence of growth and with grazing as the only loss process, the specific loss rate from species i (m_i; day^{-1}) will be given by

$$m_i = -\frac{\dot{C_i}}{C_i} = F_i Z , \tag{6.6}$$

where C_i [(mg C) l^{-1}] is the biomass of species i, Z [(mg C) l^{-1}] the grazer biomass, and F_i [l (mg C)$^{-1}$ day^{-1}] the specific clearance rate on prey species i.

Selective feeding on two cooccurring prey species ($i = 1, 2$) will then be equivalent to $m_1 \neq m_2$. If we choose indices such that species 1 is the preferred prey ($m_1 \geq m_2$) we can define a selectivity index ϕ ($0 \leq \phi \leq 1$) by

$$\phi = \frac{m_2}{m_1} = \frac{F_2}{F_1} . \tag{6.7}$$

$\phi = 0$ corresponds to complete avoidance or rejection of prey species 2 and $\phi = 1$ to perfectly nonselective feeding on the two species. It should be noted that the index ϕ is identical to selectivity indices proposed by Jacobs (1974) and Vanderploeg and Scavia (1979).

Selectivity is said to be *invariant* if the ratio of grazer-induced specific loss rates among two prey species is independent of their relative abundance (Sterner 1989). In contrast, *variant* selectivity implies frequency-dependent switching in feeding behavior, such that the most common prey species is also the most heavily predated one (Murdoch 1969). Restated in terms of Eq. (6.7), invariant selectivity implies that ϕ is constant, and thus independent of the biomasses of the prey species. By rearranging Eq. (6.7), it is easily seen that constant ϕ is equivalent to $F_2 = \phi F_1$, or, if we drop the indices, that $F_1 = F$ and $F_2 = \phi F$. The corresponding ingestion rate (I; day^{-1}) on a mixture of the two species will then be given by

$$I = F_1 C_1 + F_2 C_2 = F(C_1 + \phi C_2). \qquad (6.8)$$

Vanderploeg et al. (1984) introduced the concept of *effective food concentration* [C; (mg C) l^{-1}], which is the virtual food concentration that controls the functional response in a grazer feeding selectively on a mixed phytoplankton community. If we require that the effective food concentration (C) must satisfy $I = F C$, then Eq. (6.8) implies that C must be a linear function of the prey biomasses:

$$C = C_1 + \phi C_2. \qquad (6.9)$$

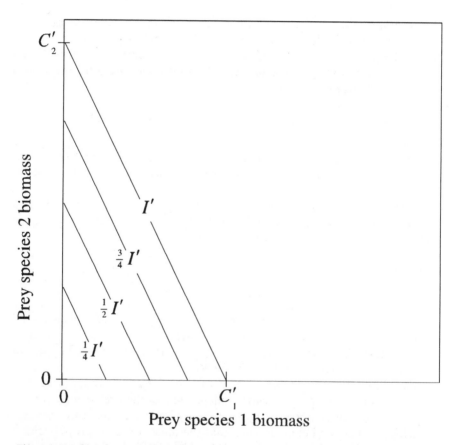

Fig. 6.6. Functional response, given by Eqs. (6.9), (6.10), of an invariant selective grazer feeding on two prey species. *Labels* denote level curves of constant specific ingestion rate

This means that isolines of the functional response will be straight lines in the C_1 - C_2 plane corresponding to $C_1 + \phi C_2$ being constant.

In accordance with Section 4.4, we can assume that ingestion is a piecewise linear function of the effective food concentration (C), such that

$$I = I' \text{Min}(1, C/C'), \tag{6.10}$$

with I' and C' being the maximal ingestion rate and the incipient limiting food level, respectively, as in Eq. (4.9). If we define C'_1 as the incipient limiting food level for grazing on species i alone, then we must have $C'_1 = C'$ and $C'_2 = \phi^{-1}C'$, which again implies that $\phi = C'_1/C'_2$. This means that the ingestion rate will be constant and equal to I' for all biomass pairs C_1 and C_2 located above the line from C_1 to C_2 in the C_1- C_2 plane (Fig. 6.6), while the isolines of the functional response will be parallel to this line. Comparing Fig. 6.6 with the resource classification scheme of Leon and Thumpson (1975) suggests that a pair of phytoplankton species subject to invariant selective grazing can be regarded as perfectly substitutable resources.

Food Selection in Daphnia. Although the issue is not entirely settled, several recent studies (DeMott 1982; Knisely and Geller 1986; Sommer 1988b) indicate that selectivity in *Daphnia* (and closely related genera) is of the invariant type. This should also be expected from the mechanics of cladoceran filter-feeding, which lacks the sophisticated mechanisms for selection of individual particles that are found, for example, in calanoid copepods (see, e.g., the review by Sterner 1989).

Knisely and Geller (1986) performed an impressive series of food selection experiments on naturally occurring phytoplankton species by two species of *Daphnia* (*D. hyalina* and *D. galeata*) over a seasonal cycle in Lake Constance. In a somewhat different approach, Sommer (1988b) studied food selection in two *Daphnia* species (*D. longispina* and *D. magna*) in a two-stage chemostat system inoculated with natural phytoplankton communities. In both studies, the selectivity index ϕ was calculated from loss rates estimated directly from decreases in cell counts. Since there was apparently low similarity between the phytoplankton communities used in the two studies, they must be considered complementary. Figure 6.7 suggests that the investigated phytoplankton genera cannot be separated into two disjunct groups of edible and inedible species, but instead form a continuum from high to low edibility. It also appears that no genus was uniformly the preferred food (mean $\phi < 1$ for all genera) and no genus was completely inedible (all genera have mean $\phi > 0$).

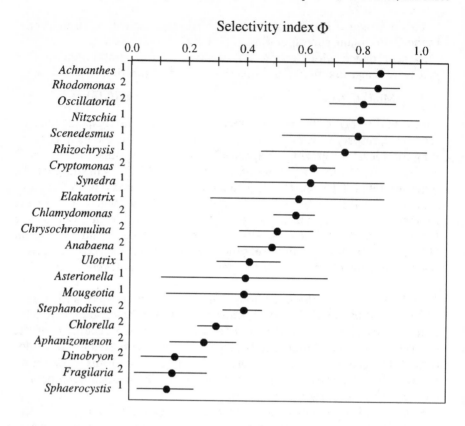

Fig. 6.7. *Daphnia* selectivity on different phytoplankton genera, expressed as the selectivity index given by Eq. (6.7) (mean ± 1 standard deviation). (*1* Data from Sommer 1988b, *2* data from Knisely and Geller 1986)

Model Equations. In order to investigate under what conditions a community with efficient harvesting of primary production can be resistant to invasion by grazing-resistant algae, we can build directly on the model developed in the previous section [Eqs. (6.2)-(6.5)]. If we consider a system with two phytoplankton species that can have differential loss rates, the mass-balance equations (6.2)-(6.5) need to be generalized to

$$\dot{C}_i = \left(\mu_i - \left(D + \sigma_i + F_i Z\right)\right)C_i \tag{6.11}$$

$$\dot{P}_i = \nu_i C_i - \left(D + \sigma_i + F_i Z\right)P_i, \tag{6.12}$$

where σ_i and F_i are the sinking loss and clearance rate of species i ($i = 1, 2$), and all other symbols as in Eqs. (6.2)-(6.3). Following the formalism of invariant selective grazing, we assume that $F_1 = F$ (i.e., species 1 is the preferred species) and $F_2 = \phi F$, with ϕ being the selectivity on species 1 with respect to species 2. If we introduce the effective food concentration C, defined by Eq. (6.9), we will have that $F = I/C$ with the ingestion rate I given by the piecewise linear function [Eq. (6.10)]. As cells are ingested as whole entities, we can define the effective concentration of algal P in analogy with Eq. (6.9) as

$$P = P_1 + \phi P_2. \qquad (6.13)$$

The remaining mass-balance equations for zooplankton carbon [Eq. (6.4)] and dissolved inorganic phosphorus [Eq. (6.5)] will be unaffected by redefining the symbols C and P [by Eqs. (6.9) and (6.13)] from total concentrations to effective concentrations of algal carbon and phosphorus. In fact, the present model will be identical to the one in the previous section when $\phi = 1$.

Edibility and Invadability in a Eutrophication Gradient. As in the previous section, the condition for coexistence of two phytoplankton species with different edibilities at a fixed equilibrium point is that all derivatives in the mass-balance equations vanish. Setting $\dot{C}_i = 0$ and $\dot{P}_i = 0$ in Eqs. (6.11) and (6.12), and eliminating the loss terms, implies that $v_i = \mu_i Q_i$ for $i = 1, 2$. In other words, uptake and growth must balance in both species, such that both species must satisfy their respective Monod relationships (cf. Section 3.3). Equation (6.11) implies that $\mu_1 = D + \sigma_1 + FZ$ and $\mu_2 = D + \sigma_2 + \phi FZ$ at the equilibrium point. Eliminating the grazing term (FZ) gives the following relationship between the equilibrium growth rates for a given value of the selectivity index ϕ

$$\phi = \frac{\mu_2 - (D + \sigma_2)}{\mu_1 - (D + \sigma_1)}. \qquad (6.14)$$

If we substitute the Monod equations (3.10) for μ_1 and μ_2 into Eq. (6.14), we find that the equilibrium condition [Eq. (6.14)] is equivalent to a quadratic equation in terms of the dissolved inorganic P concentration [S; (μg P) l^{-1}]. In other words, coexistence of the two species at fixed densities is only possible for (at most) two specific values of the equilibrium dissolved inorganic P concentration, corresponding to the roots of this quadratic equation. For all other combinations of equilibrium growth rates (μ_1 and μ_2), species 2 will be able to invade a species 1 equilibrium and competitively exclude species 1 whenever i. e.,

$$\mu_2 - (D + \sigma_2) > \phi (\mu_1 - (D + \sigma_1)).$$

Equation (6.14) implies that if species 2 is competitively superior in the absence of grazing [that is, if $\mu_2 - (D + \sigma_2) > \mu_1 - (D + \sigma_1)$ for all $S > 0$], then species 2 will also be competitively superior in the presence of grazing (that is, for all $0 \le \phi \le 1$). This means that for the resident species to have any chance at all of resisting invasion by less edible species, we must assume a tradeoff between competitive ability and predator defence. The literature data compiled in Table A10.1 showed indications of such a tradeoff in the sense that species classified as resistant to grazing either from cell or colony dimensions or by the presence of a digestion-resistant cell envelope had a significantly lower maximum growth rate. On a theoretical basis, several recent models (Grover 1989; Aksnes and Egge 1991) have predicted that large, grazing-resistant species should be inferior competitors for nutrients, although such a tendency was not supported by the literature survey of phosphorus uptake affinities in Table A10.1.

If we instead take ϕ as the dependent variable, we can interpret the right-hand side of Eq. (6.14) as the critical selectivity ϕ^*, where a competitive shift occurs for a given combination of equilibrium growth rates (μ_1 and μ_2). Since the critical selectivity ϕ^* is the ratio of the net growth rates in the absence of grazing when both species grow according to the Monod model, it can be expressed as a function of the equilibrium inorganic P concentration (S). Figure 6.8 illustrates the utility of this concept for two model

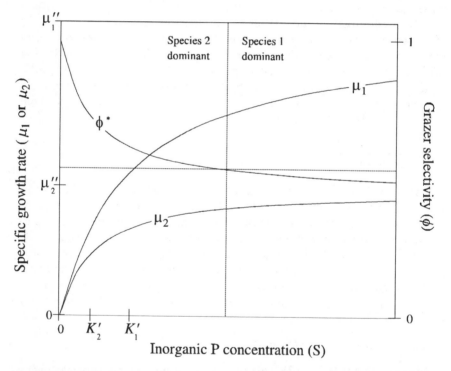

Fig. 6.8. Critical selectivity for a pair of competing phytoplankton species as function of DIP concentration. *Solid curves* labeled μ_1 and μ_2 are Monod curves (left axis) for the preferred species (1) and the grazing-resistant species (2). *Solid curve* labeled ϕ is the critical selectivity (right axis) of the grazer for species 2 relative to species 1, given by Eq. (6.14). For a given grazer selectivity (*dashed horizontal line*) a dominance shift between the two species will take place at the DIP concentration marked by the *vertical dashed line*

species constructed by making them identical with respect to all growth and uptake parameters except the maximal growth rate, which is assumed to be highest in species 1 ($\mu''_1 > \mu''_2$). Since the initial slopes of the Monod curves [given by Eq. (3.16)] will be identical for this particular choice of parameters, the Monod curves will be very close to each other for S close to zero, and diverge progressively with increasing S. This explains why ϕ^* approaches 1 for $S \to 0$ and decreases towards an asymptotic level $\approx \mu''_2/\mu''_1$ for $S \to \infty$. For a grazer with a given selectivity ϕ for species 2 (dashed horizontal line in Fig. 6.8), a shift from dominance by species 2 to species 1 will take place at the critical inorganic P concentration corresponding to $\phi^* = \phi$ (dashed vertical line in Fig. 6.8).

In the following we will investigate how different kinds of competitive handicaps can affect the ability of a grazing-resistant species to invade a simple two-species system consisting of a grazer and its preferred prey. As in the previous Section, we can perform simulated invasion experiments by first letting the system converge to its stationary state (a stable focus or a limit cycle) with only species 1 resident. The invasion is then classified as successful if species 2 is able to increase its biomass from an initial inoculum, arbitrarily set to 1% of the initial species 1 biomass.

By making repeated invasion experiments under different phosphorus supply rates [$L_P = DP_L$; (μg P) l^{-1} day^{-1}], we can express the vulnerability to invasion as function of the external forcing variable L_P instead of the internal state variable S. As in Section 6.1, we can track the two major dynamic modes of the target system with only species 1 resident by either slowly increasing L_P from an initial low value (tracking the stable focus) or slowly decreasing L_P from an initial high value (tracking the limit cycle). By repeated simulations with different values of ϕ we can find numerically the critical selectivity ϕ^* for a given P supply rate (L_P). The search for the critical selectivity ϕ^* is efficiently implemented as a bracketing and bisection procedure similar to those described in Appendix A9.

We will first consider the case where the invading species has no grazing-independent disadvantage in terms of differential nutrient efflux or differential sinking loss. That is, we will assume that $S'_1 = S'_2 = 0$ (no threshold for positive net uptake), and that $\sigma_1 = \sigma_2 = 0.0008$ day^{-1} (the nominal sinking loss rate used in previous Sections). Figure 6.9 shows three different combinations of growth and uptake parameters in the two competing phytoplankton species, corresponding to decreasing competitive ability of the invading species (2) with respect to the resident species (1). The maximal growth rates and maximal P uptake affinities are varied according to Table 6.1, while the phosphorus subsistence quotas and storage capacities are in all cases assumed identical to the "typical" values proposed in Table 3.1 [$Q'_1 = Q'_2 = 3.8$ (μg P) (mg C)$^{-1}$ and $Q''_1 = Q''_2 = 28.5$ (μg P) (mg C)$^{-1}$].

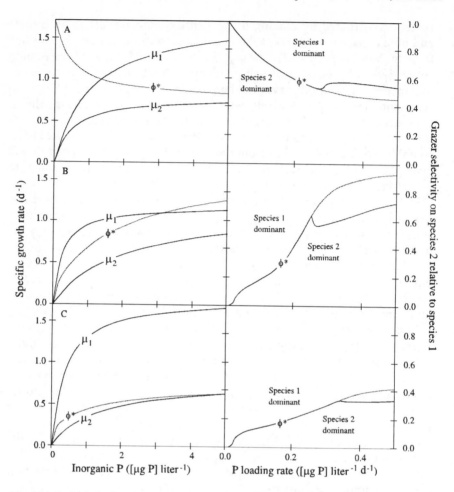

Fig. 6.9A-C. Critical selectivities for different pairs of competing phytoplankton species (**A, B,** and **C**) in an eutrophication gradient. *Left panels* are equivalent to Fig. 6.8, showing Monod curves for the resident species (μ_1) and the invading species (μ_2), as well as the resulting critical selectivities ϕ^* given by Eq. (6.14). *Right panels* show the numerically determined relationships between ϕ^* and the external P supply rate (at a constant dilution rate $D = 0.01$ day^{-1}). *Dashed curves in the right panels* represent the continuations of the ϕ^* curve at the stable focus, above the bifurcation point

Case A in Fig. 6.9 is identical to the species pair considered in Fig. 6.8, with the resident and the invading species differing only in their maximal growth rates. As discussed above, this situation gives increasing advantage to species 1 (in terms of decreasing ϕ^*) with increasing equilibrium dissolved inorganic P concentration (*S*). This pattern is repeated when ϕ is expressed as function of P loading (L_p), such that for this particular case the invader must have an increasingly more efficient predator defence in order

Table 6.1. Combinations of maximal growth rates and maximal P uptake affinities used in the species pairs considered in cases A, B, and C of Fig. 6.9, as well as the corresponding Monod half-saturation parameters [given by Eq. (3.11)].

	Parameter	Case A	Case B	Case C	Unit
μ''_1	Maximal growth rate	1.80	1.20	1.80	day^{-1}
μ''_2	Maximal growth rate	0.80	1.20	0.80	day^{-1}
α'_1	Maximal P uptake affinity	6.5	18.2	18.2	l (mg C)$^{-1}$ day^{-1}
α'_2	Maximal P uptake affinity	6.5	2.3	2.3	l (mg C)$^{-1}$ day^{-1}
K'_1	Monod parameter	1.05	0.25	0.38	(μg P) l^{-1}
K'_2	Monod parameter	0.47	1.98	1.32	(μg P) l^{-1}

to be the superior competitor when the P loading rate is increased. The nonlinear relationship between equilibrium inorganic P and P loading (cf. Fig. 6.1) causes a nonlinearity in the mapping from ϕ^* as function of S (left panel in Fig. 6.9) to ϕ^* as a function of L_p (right panel in Fig. 6.9). This distortion makes ϕ^* decrease more slowly from 1 at low loading rates and more steeply towards the asymptotic level $(\mu''_2 - (D + \sigma_2))/(\mu''_1 - (D + \sigma_1))$ at high loading rates. Since species 1 is more vulnerable to invasion at low growth rates, and since the average growth rate is reduced in the limit cycle compared to the stable focus, the onset of limit cycling behavior at the bifurcation point makes the system less resistant to invasion.

Case B in Fig. 6.9 explores a situation where species 1 and 2 differ only with respect to the maximal P uptake affinities, which are assigned the high and low values proposed in Table 3.1. This situation could result if large, grazing-resistant cells have less nutrient-transport sites per cell volume, or if a digestion-resistant cell envelope also acts as a diffusion-barrier for nutrient transport. In contrast to case A, the competitive advantage of species 1 will now decrease with increasing S or L_p, since both species approach the same maximal growth rate (that is, $\phi^* \to 1$ as $S \to \infty$). The nonlinear relationship between the gradient axes in the left and right parts of Fig. 6.9 becomes more pronounced in case B, where the ϕ^*-curve in the loading gradient becomes sigmoidal and more closely approaching the asymptotic value at the highest loading rate. The increasing competitive advantage of species 2 with increasing P loading also reverses the situation from case A, in that the limit cycling mode now becomes more resistant to invasion than the stable focus.

Case C in Fig. 6.9 is a combination of cases A and B in that the invading species is assumed to have a double competitive disadvantage with respect to the resident species, in terms of both maximal growth rate and maximal P uptake affinity. The resulting ϕ^*-curves resemble those in case B, but with an asymptotic level equal to $(\mu''_2 - (D + \sigma_2))/(\mu''_1 - (D + \sigma_1))$, as in case A, instead of 1, as in case B. Balancing the competitive advantage of species 1 between high and low nutrient levels gives less difference between the limit cycle and the stable focus than in the other cases.

Case D in Fig. 6.10 shows results of giving the invading species an additional, grazing-independent loss rate in terms of a nonzero threshold concentration for positive, net P uptake (that is, $S'_2 > 0$). As can be seen from equation (3.10), increasing S' will translate the Monod curve to the right, giving an intersection with the S-axis at $S = S'$. If we use $S'_2 = 0.1$ (µg P) l^{-1}, we would be in the lower end of the parameter range indicated in Section 3.4 (and well below the detection limit of standard phosphate analysis methods). If we otherwise use the same parameters as in case C, the resulting Monod curves for species 2 in cases C and D would be very hard to distinguish within normal experimental errors. On the other hand, the corresponding ϕ-curves as functions of P loading differ quite dramatically from case C to case D, where even a completely inedible species (that is, one with $\phi = 0$) will be unable to invade the system over an extended range of the eutrophication gradient.

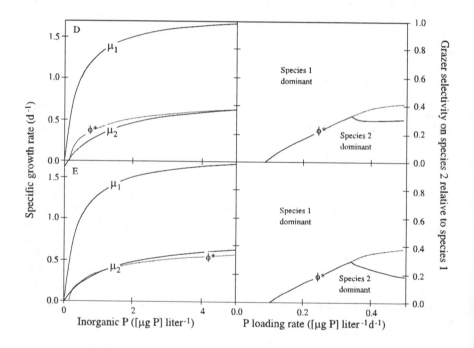

Fig. 6.10D,E. Critical selectivities for different pairs of competing phytoplankton species (D and E) in an eutrophication gradient. (See legend to Fig. 6.9)

Case E in Fig. 6.10 illustrates the effects of letting the invading species suffer from an additional competitive disadvantage in terms of increased sinking loss rate ($\sigma_1 < \sigma_2$). While nutrients lost by the efflux process considered in case D will be immediately available for the competitor, nutrients lost by sinking will be equally unavailable to all members of the community. In order to compare the two types of differential loss processes, we can, for example, assume that the invading species in both cases D and E have equal loss rates at zero inorganic P concentration. By setting $S = 0$ in Eq. (3.10), we find that this will be the case if we assume species 2 to have a sinking loss rate $\sigma_2 = 0.066$ day^{-1}, which will be in the lower half of the range considered as typical for large, nonmotile algae in Section 2.3. If we keep all other parameters identical to those in case C; the Monod curves will also be identical in cases C and E, while the ϕ^*-curve as function of S will be similar to the one from case D, but with a slightly lower asymptote due to the contribution from σ_2 in the nominator of Eq. (6.14). Up to the bifurcation point, the ϕ^*-curves as functions of P loading are nearly identical for cases D and E. After the onset of limit cycling behavior, the two cases differ in that the ϕ^*-curves in case E starts to decrease while it remains at a rather constant level in case D. This difference between the two loss processes might be caused by an increasing sinking loss of phosphorus from the species 2 compartment in the undergrazed phase of the limit cycle.

The cases illustrated in Figs. 6.9 and 6.10 indicate that for a grazer-controlled plankton community to be resistant to invasion by phytoplankton species with a grazing refuge, the resident species must have higher capacities for both growth and nutrient uptake. Superiority in only one of these aspects will leave the system open to invasion by species with only a modest level of predator defence at either the low or the high end of the nutrient loading gradient. Resistance to invasion by completely inedible phytoplankton species seems possible only if this level of predator defence can only be realized at the expense of increased loss rates due to grazing-independent processes. If the completely inedible species is capable of attaining positive net growth at all [that is, if $\mu''_2 - (D + \sigma_2)$], then it will nevertheless be able to invade at a sufficiently high P loading rate.

The takeover by species 2 was always absolute for selectivities below ϕ^*, with no intermittent interval allowing coexistence between the two species. It thus appears that the activity of an invariant selective grazer does not promote the coexistence of its prey species, like the non-selective grazing considered in Section 6.1. This also means that the present model is unable to represent a situation with stable coexistence of large, inedible "canopy" species and small, edible "undergrowth" species identified as step 8 in the PEG model of plankton succession (Sommer et al. 1986).

6.3 Differential Nutrient Recycling and Resource Supply Ratios

Up to this point, we have only considered systems with one limiting resource. Even if phosphorus is generally identified as the element most likely to become limiting to phytoplankton growth in lakes, algal biomass contains many other essential elements that also must be present in sufficient quantities for the maintenance of vital cellular functions. Differential requirements for essential elements like silicon and nitrogen among the major phytoplankton groups are thought to be important in determining major events in phytoplankton succession (cf. steps 11, 12, and 13 in the PEG model of plankton succession; Sommer et al. 1986).

The resource competition model of Tilman (1980) has generally been successful in predicting the outcome of competition experiments between diatom species in gradients of Si:P supply ratios (summarized in Table 3.1 in Sommer 1989b). This predictive success implies a strong tradeoff between competitive abilities for Si and P in different diatom species, thus excluding the possible existence of a "superspecies" with superior competitive abilities for both elements. Based on the variability in "optimum" N:P ratios, introduced by Rhee and Gotham (1980), in different phytoplankton species, a similar tradeoff between competitive ability for N and P has been proposed. The "optimum" N:P ratio is defined as the ratio between the subsistence quotas (Q') for N- and P-limited growth according to the Droop model [Eq. (3.2)]. Although the "optimum" N:P ratio has been interpreted as the N:P supply ratio leading to equal N- and P-limitation, this interpretation has been criticized by several authors, as summarized by Turpin (1988). As shown in Section 3.7, the optimum N:P supply ratio will actually be growth rate-dependent, unless the asymptotic growth rate parameters (μ') of the Droop model are identical for N- and P-limited growth [which will be strictly true only if the storage capacities (Q''/Q') for N and P are equal; cf. Eq. (3.3)].

Zooplankton grazers release the remains of ingested diatom frustules mostly as particulate silicon (Paasche 1980). The limited recycling of Si by grazers will therefore decrease the Si:P supply ratio to the disadvantage of diatoms in the presence of herbivores (Sommer 1988b). In contrast, both N and P are found to be recycled by herbivores in forms easily available to algae (Lehman 1984). Since herbivores need both N and P for their production of new biomass, the N:P ratio of recycled nutrients from grazing zooplankton will be determined by the N:P ratio of the food in relation to the requirements of the grazer (Sterner 1990; Andersen and Hessen 1991).

If the interplay between N:P ratios of food items and recycled nutrients leads to a fluctuating N:P supply ratio, then zooplankton grazing could, according to the theory of Tilman (1982), potentially mediate the coexistence of more than two phytoplankton species on two limiting resources, thus violating the competitive exclusion principle in the strictest sense. If, instead, zooplankton grazing leads to N:P supply ratios that diverge from those of the external supply, as suggested by Sterner (1990), the presence of grazers would tend to destabilize pairs of phytoplankton species that would otherwise coexist for certain N:P supply ratios in the absence of grazing. In order to investigate these two possible outcomes of competition for N and P in the presence of grazers, we will need to describe N and P utilization by zooplankton at some more detailed levels.

Differential N and P Recycling. We will assume that grazers maintain an internal homeostasis leading to constant C:N:P ratios of grazer biomass, as indicated by the data of Andersen and Hessen (1991). In Section 4.5 it was shown that observed P release rates would be consistent with the maintenance of a constant animal P:C ratio if grazer growth rate was described by a piecewise linear function of food P:C ratio [Eq. (4.19)]. Since this growth model is equivalent to a Liebig-type minimum law, a threshold model seems to be appropriate for describing grazer growth on a food source that can be both N- and P-deficient. If, in analogy with Eq. (4.25), we define the grazer growth rate on N- and P-sufficient food as g' (day^{-1}), then grazer growth rate (g; day^{-1}) as function of food N- and P-content can be written as

$$g = g' \mathrm{Min}(1, Q_P/\theta_P, Q_N/\theta_N), \tag{6.15}$$

where Q_P and θ_P are the P:C ratios [(μg P) (mg C)$^{-1}$], and Q_N and θ_N are the N:C ratios [(μg N) (mg C)$^{-1}$] of food particles and grazers, respectively. In accordance with Eq. (4.21), the release rate of element R [ρ_R; (μg R) (mg C)$^{-1}$ day^{-1}, with R = N or P] will be given by the difference between ingestion and utilization:

$$\rho_R = I Q_R - g\theta_R, \tag{6.16}$$

with I being the specific food carbon ingestion rate (day^{-1}). If we substitute Eq. (6.15) into Eq. (6.16), the N:P ratio of recycled nutrients becomes

$$\frac{\rho_N}{\rho_P} = \frac{Q_N - K_1' \mathrm{Min}(1, Q_P/\theta_P, Q_N/\theta_N)\theta_N}{Q_P - K_1' \mathrm{Min}(1, Q_P/\theta_P, Q_N/\theta_N)\theta_P}, \tag{6.17}$$

where we have introduced the dimensionless gross growth efficiency $K_1' = g'/I$. By rearranging Eq. (6.17), we can express the N:P ratio of recycled nutrients relative to the food N:P ratio in terms of the relative N- and P-contents of the food (Q_N/θ_N and Q_P/θ_P):

$$\frac{\rho_N/\rho_P}{Q_N/Q_P} = \frac{1 - K_1' \mathrm{Min}(1, Q_P/\theta_P, Q_N/\theta_N)(Q_N/\theta_N)^{-1}}{1 - K_1' \mathrm{Min}(1, Q_P/\theta_P, Q_N/\theta_N)(Q_P/\theta_P)^{-1}}. \tag{6.18}$$

The right-hand side of Eq. (6.18) will be >1 whenever $Q_N/Q_P > \theta_N/\theta_P$, and <1 when $Q_N/Q_P < \theta_N/\theta_P$. That is, when the N:P ratio of the food is higher than the requirements of the grazer, the N:P ratio of recycled nutrients will be even higher, and vice versa. If the recycled nutrients are completely assimilated by the food organisms, then the food N:P ratio will either increase or decrease, except for the special case when food composition exactly matches the requirements of the grazer. It should be noticed that this divergence of N:P ratios as a result of grazing is not dependent on the particular function chosen in Eq. (6.15); it would be the same for a general growth model $g = G(Q_P/\theta_P, Q_N/\theta_N)$, with G being an arbitrary bivariate function.

The function (6.18) can be visualized by observing that the right-hand side is dependent only on the carbon-relative food contents of N and P (Q_N/θ_N and Q_P/θ_P). Figure 6.11 shows isolines of the surface described by Eq. (6.18) for the particular case of a gross growth efficiency $K'_1 = 0.5$. All isolines with $(\rho_N/\rho_P)/(Q_N/Q_P) < 1$ are located below the diagonal line corresponding to optimal food composition ($Q_N/\theta_N = Q_P/\theta_P$), while all isolines > 1

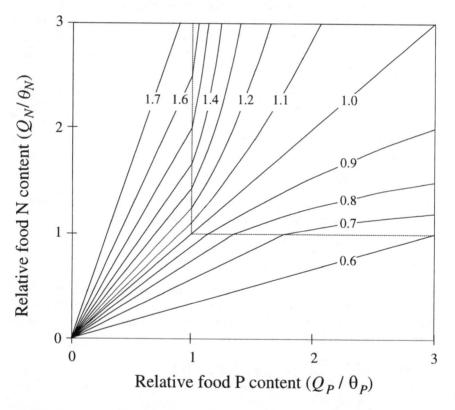

Fig. 6.11. N:P ratio of recycled nutrients relative to food N:P ratio as function of relative food N and P content. *Labeled curves* are isolines of Eq. (6.18), with a gross growth efficiency of 0.5

are above the diagonal. The isolines are seen to diverge when food elemental composition becomes non-limiting ($Q_N > \theta_N$ and $Q_P > \theta_P$); that is, the effect of differential nutrient utilization by grazers on N:P ratios of recycled nutrients diminishes when the food contains a surplus of both elements.

Competition in an N:P Loading Gradient. We have now reached the point where we can assemble the submodels of phytoplankton competition and zooplankton recycling into a system driven by the external supply rates of N and P. In order to do so, we must first rewrite the phytoplankton mass-balance equations (3.18), (3.19) to include the loss rate due to grazing. The addition of a non-selective grazer, as discussed in Section 6.1, to the system, gives the modified mass-balance equations as

$$\dot{R}_i = v_{R.i} C_i - (D + F Z) R_i \qquad (6.19)$$

$$\dot{C}_i = (\mu_i - (D + F Z)) C_i \qquad (6.20)$$

for $i = 1, 2$ and R = N, P (that is, a total of six differential equations). As in Eqs. (6.2), (6.3), Z is grazer biomass [(mg C) l^{-1}] and F is the clearance rate of the grazers [1 (mg C)$^{-1}$ day^{-1}]; otherwise the notation is unchanged from Eqs. (3.18), (3.19). If we, as in Section 6.1, denote the concentrations of food C, N, and P that are available to the grazers by $C = C_1 + C_2, N = N_1 + N_2$, and $P = P_1 + P_2$, the P:C and N:C ratios of available food become $Q_P = P/C$ and $Q_N = N/C$. With this notation, we can directly use the expressions (6.15), (6.16) for zooplankton growth and nutrient release as function of food composition. Thus, the modified mass-balance equations for the dissolved inorganic nutrient pools can be written as:

$$\dot{S}_R = \rho_R Z + D(R_L - S_R) - \sum_i v_{R.i} C_i \qquad (6.21)$$

for R = N, P. The first term in Eq. (6.21) is the zooplankton release rate of element R, given by Eq. (6.16), while the remaining terms describing loading, dilution loss, and uptake are the same as in Eq. (6.29). Finally, the zooplankton biomass mass-balance equation will be the same as Eq. (6.4), but with the growth rate given by the threshold model [Eq. (6.15)]:

$$\dot{Z} = (g - (D + \delta)) Z \qquad (6.22)$$

All phytoplankton parameters will be assumed to be the same as those in Table 3.4, which were deliberately chosen to conform with the conditions for coexistence from Appendix A1 (as illustrated in Fig. 3.9). With the exception of the grazer N:C ratio (θ_N), all zooplankton parameters will be identical to those in Table 5.1. Using a typical N:C ratio for *Daphnia*, as reported by Andersen and Hessen (1991), θ_N can be set to 190 (µg N) (mg C)$^{-1}$.

Mapping the N:P Loading Gradient. If we choose a constant dilution rate ($D = 0.01$ day^{-1}, as in the foregoing simulations), we can construct a gradient of N:P loading ratios by varying the input concentrations of N and P (N_1 and P_1). The competitive abilities of the two species at different external N:P supply ratios can then be studied by simulated reciprocal invasion experiments, as in Section 6.2. Each set of invasion experiments was started with a simulation run designed to establish one of the species as the resident.

Species 1 was established as resident at an N:P loading ratio sufficiently high to ensure P-limitation in the stationary state ($N_1 : P_1 = 15$; cf. Fig. 3.10), while species 2 was established as resident at a low N:P loading ratio ($N_1 : P_1 = 5$; cf. Fig. 3.10) to ensure N-limitation. As in the previous Section, initial conditions for the initial runs were randomly chosen, but with all state variables for the invading species set to zero. Each invasion experiment in the same set used an initial condition based on the stationary state with only the resident species present, with the addition of a small inoculum of the invading species (equal to 1% of resident biomass). The outcome of the invasion was classified as exclusion of species i if final species i biomass was <1% of total phytoplankton biomass, and as coexistence if both species had >1% of final biomass.

By performing a set of invasion experiments at a fixed P loading rate, and varying the N loading rate, the intervals of N supply rates leading to exclusion or coexistence can be efficiently found by a binary search procedure similar to the one used in Section 6.2. Repeating this process for a range of P loading rates mapped out the regions of exclusion and coexistence in the two-dimensional gradient of N and P loading rates. By slowly increasing the P loading from a low value, or by decreasing it from an initial high value, the two major dynamic modes of the system (the stable focus and the limit cycle) can be tracked individually, as in previous sections.

Using either species 1 or species 2 as resident gave practically the same results in the simulated invasion experiments: an abrupt shift from competitive dominance by one species to the other taking place at critical N:P loading ratio equal to the N:P ratio of the grazer ($N_L/P_L = \theta_N/\theta_P$), leading to the exclusion of species 2 at high N:P ratios ($N_L/P_L > \theta_N/\theta_P$) and species 1 at low N:P ratios ($N_L/P_L < \theta_N/\theta_P$). The positive feedback generated by differential nutrient recycling apparently makes N:P supply ratios diverge to the extent that coexistence becomes impossible, even for small deviations from the optimal N:P supply ratio relative to grazer requirements (that is, $N_L/P_L = \theta_N/\theta_P$). Even if some kind of evolutionary design tradeoff could make species able to coexist at certain N : P supply ratios by conforming to the conditions illustrated in Fig. 3.9 (which by itself might seem quite unlikely), such a situation of competitive balance would be very sensitive to the effects of differential nutrient utilization by zooplankton grazers. This makes it less likely that the explanation to the paradox of the plankton should be found in the theory of competition for two or more essential resources.

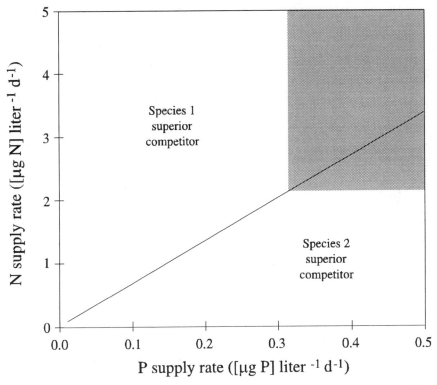

Fig. 6.12. Regions of competitive dominance of species 1 and 2 in a two-dimensional gradient of N and P loading rates (dilution rate $D = 0.01$ day^{-1}). *Shaded area* is the range of loading rates leading to a persistent limit cycle

The final outcome of the invasion experiments did not seem to be influenced by whether the resident system was in a steady-state or oscillating, although the rate with which the resident species was excluded differed markedly between these two dynamic modes. When the resident system was at a stable focus, the takeover by the invader was generally fast, with the resident species being replaced in the first 50 days of the simulation. The exclusion rate was much slower for systems with the resident species in a limit cycle (the shaded area in Fig. 6.12), often with the resident being still present after several 100 simulation days. It is reasonable to expect that on a time scale of normal seasonal succession, exclusion rates of this magnitude could easily be perceived as coexistence.

Some indication of the mechanism behind the slow exclusion rate in the limit cycle can be given by considering the phase diagram of the total N and P supply rates [computed as the first two terms of Eq. (6.21)]. Figure 6.13 shows that the N:P supply ratio fluctuates asymmetrically around the external N:P ratio, taking excursions to very high supply ratios, but also staying below the external supply ratio for a substantial part of the cycle.

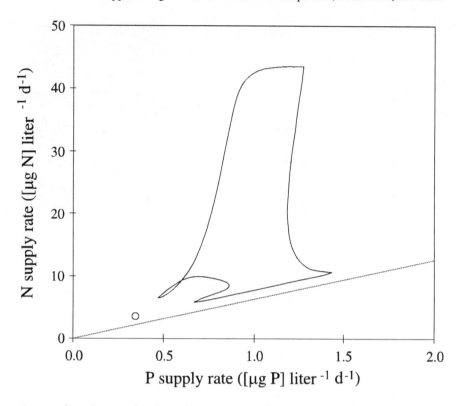

Fig. 6.13. Phase diagram of total N and P supply rates (from loading and recycling) for a system with species 2 resident in a limit cycle at external N and P loading rates of 3.5 (µg N) l^{-1} day^{-1} and 0.35 (µg P) l^{-1} day^{-1} (*open circle*); *dashed line* grazer N:P ratio

This temporal variability in the nutrient supply regime of the limit cycle apparently gives a fluctuating competitive advantage for the invading species, which confines the process of competitive exclusion to only a part of the cycle and thus reduces the overall rate of exclusion. The strong asymmetry of the fluctuating supply ratio is still too much in favor of the invading species to allow coexistence. Thus, somewhat in contrast to the theory of Tilman (1982), fluctuating resource supply ratios generated by the limit cycle do not seem to promote coexistence in the strictest sense.

   The exact coincidence of the critical loading ratio for a shift in the competitive ability of the two species with the grazer N:P ratio suggests that it is the loading ratio in relation to grazer elemental composition which determines whether N or P is to become limiting to phytoplankton growth. This simple theory might be confounded by the fact that, more or less accidentally, the *Daphnia* N:P ratio used in the model $[\theta_N/\theta_P = 6.3$ (µg N) (µg P)$^{-1}]$ is located within the interval of N:P supply ratios where coexistence is

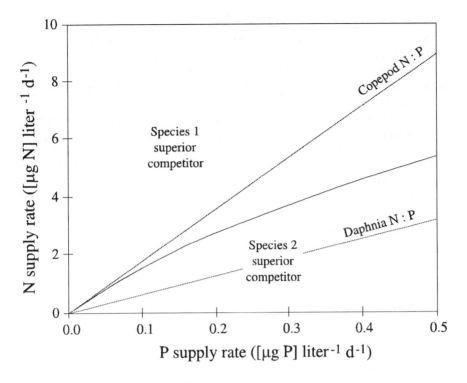

Fig. 6.14. Regions of competitive dominance of species 1 and 2 (*separated by the solid line*) in a two-dimensional gradient of N and P loading rates (dilution rate $D = 0.01$ day^{-1}). *Dashed lines* represent the N:P ratio of the copepod grazer, and, for comparison, the N:P ratio of the daphnid grazer used in Fig. 6.12

possible in the absence of grazers (cf. Fig. 3.10). To test this hypothesis, we can modify the model by assuming a grazer with a different N:P ratio, while leaving all other parameters unchanged. In the investigation of Andersen and Hessen (1991), copepods and daphnids represented the extremes with respect to N:P ratios, with copepods having more N and less P per unit dry weight. The results of running the model with a grazer elemental composition considered typical for copepods [$\theta_N = 215$ (μg N) (mg C)$^{-1}$ and $\theta_P = 12$ (μg N) (mg C)$^{-1}$; $\theta_N/\theta_P = 17.9$ (μg N) (μg P)$^{-1}$] are shown in Fig. 6.14.

Changing the N:P ratio of the grazer preserves the qualitative result that the boundary between the regions of competitive dominance is sharp, with a negligible intermittent region of coexistence. While the critical N:P loading ratio lies close to the copepod N:P ratio at low loading rates, it deviates markedly from this line with increasing nutrient loading (Fig. 6.14). The change in grazer N and P requirements thus improves the competitive ability of species 2, but the advantage is seen to decrease with increasing nutrient loading. Qualitatively, the results in Fig. 6.14 are still in agreement with proposal of Andersen and Hessen (1991), that a change in zooplankton species composition, from copepods to daphnids, could lead to a shift from N- to P-limited phytoplankton growth, or vice versa.

While this theory finds support in a study by Elser et al. (1988), who found clear evidence for a shift from N- to P-limitation following a change from copepod to daphnid dominance in the zooplankton community, it seems likely that such a mechanism can apply only for a restricted range of N:P loading ratios. If, for lack of better data, we assume that there should be a relationship between N:P loading ratios and ratios of standing stocks of total N and P, we find in the regional eutrophication survey of Faafeng et al. (1990) that only 3 / 351 of the lakes had total N:total P < 6.3 and 24 / 351 of the lakes had total N:total P < 17.9. In other words, the mechanism of zooplankton-mediated N- and P-limitation should be limited to about 6% of the lakes considered in the NIVA study, although the roughness of this calculation calls for considerable caution.

6.4 The Fate of Zooplankton Egesta: Carbon Cycling and Chaos

All models considered thus far have implicitly made the simplifying assumption that algal biomass is the sole food source of zooplankton, and that transformations of essential mineral nutrients are closely associated with the dynamics of phyto- and zooplankton populations. It has never-theless long been recognized that the metabolism of organic carbon in eco-systems can follow two major pathways: a classical grazer food chain and a parallel detritus food chain (Odum 1962, Odum and de la Cruz 1963). Since filter-feeding zooplankton appear to play a major part in both pathways, the functional distinction between grazer and detritus food chains becomes blurred in planktonic ecosystems. Most crustacean filter-feeders have been shown to both ingest and assimilate appreciable amounts of particulate detritus (Nauwerck 1963; Wetzel et al. 1972; Langeland and Reinertsen 1982; Hessen et al. 1990), and, at the same time, also produce detritus and dissolved organic matter from living biomass through egestion of non-assi-milated food and through "sloppy" feeding (Lampert 1978; Olsen et al. 1986a; Hessen et al. 1990). Repeated ingestion and egestion by zooplankton will also gradually transform particulate detritus into forms of dissolved organic carbon that are more easily available to free-living bacteria.

While the current view holds that planktonic, free-living bacteria derive much of their energy supply from algal exudates (Riemann and Sønder-gaard 1986), a recent compilation of extracellular release rates from differ-ent pelagic systems indicates that this carbon source can seldom cover more than half of the requirements for bacterial growth (Baines and Pace 1991). It therefore seems likely that a major fraction of bacterial growth must be supported by other carbon sources, like release of dissolved organic matter from the consumer levels of the pelagic food web (Azam et al. 1983).

Bacteria have traditionally been assigned the role as decomposers in the pelagic food web (e.g., Wetzel 1975), although the past decade has seen the emergence of an alternative view, where the role of bacteria as nutrient competitors also is emphasized (Azam et al. 1983; Thingstad 1987). This view is supported by the observation that bacteria have very high requirements for phosphorus compared to common plankton algae (Vadstein et al. 1988), and that bacteria have uptake affinities for P that are at least an order of magnitude higher than is found in their algal competitors (Currie and Kalff 1984; Vadstein and Olsen 1989). As pointed out by Bratbak and Thingstad (1985), this makes the relationship between planktonic algae and bacteria an ecologically interesting case of competition and commensalism; bacteria compete with algae for mineral nutrients, but are at the same time at least partly dependent on their competitors to supply them with organic carbon skeletons in the form of extracellular release products.

Similar to the minimal representation of the internal phosphorus cycle in the lower parts of pelagic food webs, presented in Fig. 2.6, we can propose a simplified flow network for the carbon cycle in freshwater plankton communities. Figure 6.15 shows a flow diagram connecting three compartments of living carbon biomass (bacteria, algae, and zooplankton) and two nonliving carbon pools (detritus and dissolved organic carbon), while Fig. 6.16 shows the complementary phosphorus flow diagram. For clarity, fluxes to and from the pool of dissolved inorganic carbon (photosynthesis and respiration) have been omitted in Fig. 6.15.

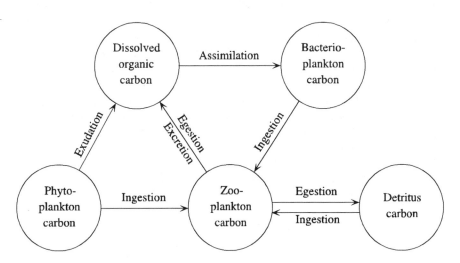

Fig. 6.15. Flow diagram illustrating major pools and pathways of the internal carbon cycle in freshwater plankton communities

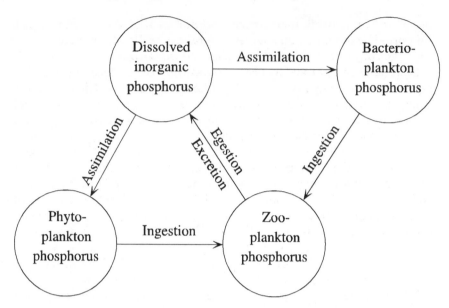

Fig. 6.16. Complementary flow diagram showing major pathways of phosphorus among the compartments identified in Fig. 6.15

Generalist filter-feeders like *Daphnia* are able to capture algae, bacteria, and detritus with comparable efficiencies (DeMott 1985; Hessen 1985b; Hessen et al. 1989). The assimilated fraction of ingested food is used for growth and maintenance, while the egested fraction enters the two nonliving carbon pools. Bacteria assimilate dissolved organic carbon from a compartment that is fed by both algal extracellular release products and zooplankton egestion and excretion. This pool of readily assimilable organic C is probably constituted by free nucleotides, amino acids, and other simple carbonyl compounds with fast turnover rates, and should therefore not be identified with the total amount of dissolved organic C (DOC) in lake water.

While total DOC is commonly found to exceed particulate organic C (POC) by an order of magnitude (Wetzel 1975), the majority of this dissolved pool is probably composed of very refractory compounds of terrestrial origin (Thurman 1985). Sunlight-induced photochemical processes have been found to transform humic substances into simple carbonyl compounds, like formate, acetate, and pyruvate, that are readily assimilable to bacteria (Kieber et al. 1989). This slow mobilization of refractory DOC can explain the observed correlation between water color and bacterial biomass in lakes (Hessen 1985a). While photochemical processes might be responsible for both the high bacterial production and the large net CO_2 flux to the atmosphere in humic lakes (Hessen et al. 1990), this mechanism is probably of minor importance in the carbon cycle of clear-water lakes. The pool of dissolved organic C in Fig. 6.15 is therefore assumed to have no external inputs.

The different biogeochemical cycles in an ecosystem are interrelated in the sense that they constitute complementary views of the same system, as pointed out by Reiners (1986). The carbon cycle in a phosphorus-limited system can therefore not be represented independently of the phosphorus cycle. The main differences between the carbon cycle in Fig. 6.15 and the phosphorus cycle in Fig. 6.21 are the reversal of the flow between the dissolved pool and the phytoplankton compartment, and the absence of a detrital pool of phosphorus. The assumption of no detrital P compartment is justified by X-ray micrographic observations of natural detritus particles (Olsen et al. 1986b). The opposite flows of dissolved organic carbon and dissolved inorganic phosphorus with respect to the phytoplankton compartment reflect the dualism as both commensalists and competitors in the interaction between algae and bacteria (Bratbak and Thingstad 1985).

Organic Carbon Release Processes. The release of extracellular products from algal photosynthesis was first recognized by Fogg (1966). Although the importance of this process was violently disputed for some time (e.g., Sharp 1977), extracellular release seems now to be generally accepted as a source of high-quality carbon to bacteria (Cole et al. 1982; Riemann and Søndergaard 1986). In laboratory studies, extracellular release has been found to increase under growth limitation (Jensen 1984), suggesting an inverse relationship where an increasing fraction of total photosynthesis is directed into extracellular release with decreasing growth rate. Inverse relationships between fraction extracellular release and total primary production or phytoplankton biomass have been found in several data sets from single habitats, but a recent compilation of measurements from different plankton communities failed to reproduce this pattern (Baines and Pace 1991).

In the 93 observations from lacustrine systems considered by Baines and Pace (1991), the fraction of extracellular release of total primary production ranged from 0.02 to 0.73, with a median of 0.21. A parsimonious model might therefore be that algal extracellular carbon release is a constant fraction $f_e = 0.21$ of total primary production. If we denote algal specific growth rate and biomass by μ_a (day^{-1}) and C_a [(mg C) l^{-1}], the total extracellular release rate from phytoplankton becomes $f_e \mu_a C_a$.

Hessen et al. (1989) found high assimilation efficiency in *Daphnia* on bacteria labeled with radioactive nucleic acids, while assimilation efficiencies by the same species were much lower on bacteria labeled with an amino acid mixture (Hessen et al. 1990). These findings are consistent if zooplankton have a differential food utilization in favor of P-rich nucleic acids compared to proteins when the bulk of available food is highly P-deficient relative to the demands for balanced growth. This view is supported by the observation (Hessen and Andersen 1990) that zooplankton are able to utilize bacterial P with high efficiency. Zooplankton assimilation efficiencies might thus be more variable among different biochemical constituents of the total ingested food than among different kinds of food particles. In a phosphorus-limited system, zooplankton growth might therefore be controlled by the total amount of ingested P in relation to ingested C.

If we assume that the grazers are able to capture algae, bacteria, and detritus with comparable efficiencies, the concentrations of food P and C available to the grazers can be written as $P = P_a + P_b$ and $C = C_a + C_b + C_d$, with P_a and P_b [(µg P) l^{-1}] being the concentrations of algal and bacterial P, and C_a, C_b, and C_d [(mg C) l^{-1}] the concentrations of algal, bacterial, and detrital C. With this notation we can apply the same zooplankton growth model as in the previous chapter (Eqs. (5.5) and (5.6)). While the phosphorus cycle is semiconservative in the sense that recycled P is generally available to autotrophs, the energy flow in food webs (which closely follows the organic carbon cycle) is necessarily dissipative, with respiratory energy losses from every biotic compartment. In contrast to P release by grazers, as described in Section 4.5, organic C release must also account for respiratory losses. If we denote the specific C release of the grazers by ρ_c (day^{-1}), the release rate can be described by the balance between ingestion (I), growth (g), and respiration (r):

$$\rho_C = I - (g + r). \qquad (6.23)$$

Equation (6.23) describes the total release rate of both particulate and dissolved forms of organic carbon, which must be further differentiated into a fraction $f_p \rho_c$ entering the pool of particulate detritus and a fraction $(1 - f_p)\rho_c$ entering the dissolved organic pool. According to Olsen et al. (1986a), the particulate fraction amounted to 80% of total organic C release (that is, $f_p = 0.8$), for *Daphnia* feeding on *Scenedesmus* under varying degrees of P limitation.

Carbon Cycling in a Eutrophication Gradient. Mass-balance equations for algae and bacteria in the presence of a nonselective grazer can be described in the same formalism as in Section 2.5. If we use indices a and b denoting process rates and state variables of algae and bacteria, respectively, the balances between gains and losses from each compartment of carbon biomass [C_a and C_b; (mg C) l^{-1}] and particulate phosphorus [P_a and P_b; (µg P) l^{-1}] can be written as

$$\dot{C}_i = \left(\mu_i - (D + FZ)\right)C_i \qquad (6.24)$$

$$\dot{P}_i = v_i C_i - (D + FZ)P_i \qquad (6.25)$$

for $i = a, b$. As before, D and FZ are losses due to dilution and grazing, while the terms μ_i and $v_i C_i$ denote growth and P uptake in compartment i. Phosphorus uptake is assumed to be controlled by internal and external P concentrations as in Eqs. (3.4), (3.5), with the parameters given in Table 3.3. Algal growth rate (μ_a; day^{-1}) is determined by internal P stores according to the Droop model [Eq. (3.2)], and also with parameters as in Table 3.3.

Bacterial growth rate (μ_b; day^{-1}) is assumed to be controlled by the availability of P and C according to a threshold model; that is, if we use symbols $\mu_{P,b}$ and $\mu_{C,b}$ for the potential P- and C-limited growth rates, then $\mu_b = \text{Min}(\mu_{P,b}, \mu_{C,b})$. The potential P-limited growth rate, $\mu_{P,b}$, is given by the Droop model [Eq. (3.2)] with the parameters in Table 3.3, as with the algal compartment. The potential C-limited growth rate, $\mu_{C,b}$, is determined by the carbon assimilation rate and the efficiency with which assimilated carbon can be converted into new biomass.

Carbon-specific growth efficiencies in bacteria have been found to be quite variable, with strong indications of modulation by the mineral nutrient supply. In nutrient-sufficient systems, maximum growth yields around 50-60% of assimilated carbon have been reported, while 40-90% of assimilated carbon might be lost as respiration in nutrient-limited systems (Azam et al. 1983). Respiratory losses can be considered as the sum of maintenance metabolism and the costs of synthesizing new biomass from simple precursors, which can be written as

$$\mu_{C,b} = \varepsilon_b u_b - r_b, \tag{6.26}$$

where u_b (day^{-1}) is the specific assimilation rate, r_b (day^{-1}) the rate of maintenance metabolism, and ε_b a dimensionless assimilation efficiency. Parameters values $\varepsilon_b = 0.5$ and $r_b = 0.5$ day^{-1}, as used in the present model, are in reasonable agreement with observed maximum growth yields in bacteria (50-60 %; Cole et al. 1982; Bjørnsen 1986), and with general allometric relationships (e.g., Peters 1983) predicting that routine metabolism should be higher in bacteria than in metazoa.

Although carbon uptake can also be represented by an internal stores model, a completely symmetrical treatment of P and C uptake would require P and C quotas to be represented on a per-cell basis, as in the model of Thingstad (1987), thus introducing a third state variable. Since C quotas should be even less variable than P quotas, the assumption of a constant C quota, and thus of uptake being controlled by the external substrate level alone, seems justified for bacterial C uptake. In order to maintain balanced growth under nonlimiting conditions, there must be an upper limit u''_b to the C uptake rate such that $\mu'_b = \varepsilon_b u''_b - r_b$ when $\mu_b = \mu'_b$. This requirement would be satisfied by assuming the uptake rate to be related to the dissolved organic C concentration [S_C; (mg C) l^{-1}] by a rectangular hyperbola similar to the Michaelis-Menten function of enzyme kinetics:

$$u_b = u''_b \frac{S_C}{K_C + S_C}. \tag{6.27}$$

Since there is a single bacterial compartment, and thus no competition for dissolved organic C in this system, the half-saturation parameter K_C for carbon uptake can be chosen somewhat arbitrarily, with a value of $K_C = 0.1$ (mg C) l^{-1} being used in the simulations. If P is limiting, then the uptake rate given by

Eq. (6.27) would be in excess of the needs for balanced growth. In the model, this situation is dealt with by simply setting $u_b = \varepsilon^{-1}{}_b (\mu_b + r_b)$ whenever $\mu_b = \mu_{P,b} < \mu_{C,b}$.

The dynamics of the bacterial substrate pool (S_C) is determined by the balance between release from algae and zooplankton, and losses due to uptake and dilution:

$$\dot{S}_C = f_e \mu_a C_a + \left(1 - f_D\right)\rho_C Z - u_b C_b - D S_C. \tag{6.28}$$

The rate of change in the other dissolved pool of the system, dissolved inorganic phosphorus $[S_p; (\mu g\ P)\ l^{-1}]$, will be given by the balance between gains from the input loading $[L_p = DP_T; (\mu g\ P)\ l^{-1}\ day^{-1}]$ and from zooplankton release, and losses through dilution and uptake:

$$\dot{S}_p = D\left(P_L - S_p\right) + \rho_P Z - \left(v_a C_a + v_b C_b\right). \tag{6.29}$$

The final nonliving compartment of the system, detritus carbon $[C_d; (mg\ C)\ l^{-1}]$, receives its input from the particulate fraction of zooplankton egesta, and loses material through grazing and dilution:

$$\dot{C}_d = f_D \rho_C Z - \left(D + FZ\right)C_d. \tag{6.30}$$

Finally, the zooplankton biomass mass-balance equation will be the same as in the previous section, giving net growth as the difference between gross growth and losses due to dilution and mortality.

$$\dot{Z} = \left(g - \left(D + \delta\right)\right)Z. \tag{6.31}$$

In the models where the food source to the zooplankton is composed of algae competing for the same inorganic nutrient, all but one phytoplankton species could be excluded without destroying the integrity of the food chain. Due to the mutual dependencies between the compartments of the carbon cycle, we would expect all state variables of the present model to be positive as long as the grazer population is persistent. Although an analytical steady-state solution to the Eqs. (6.24), (6.25), (6.28)-(6.31) can be be found, numerical experiments indicate that this solution is dynamically unstable, except for a limited range at very low phosphorus loading rates. For obvious reasons, no attempt has been made to formally analyze the local stability properties of this system through the eigenvalues of the resulting 8×8 Jacobian matrix.

Suitable initial conditions for numerical solution of the system (6.24), (6.25), (6.28)-(6.31) can be generated by assigning random values to the state variables, subject to the constraints that all state variables must be positive, with algal and bacterial cell quotas such that $Q'_a < Q_a < Q''_a$ and $Q'_b < Q_b < Q''_b$, and with total P equal to the input concentration ($P_L = S_p + P_a + P_b + \theta Z$). Running simulations from random initial conditions for a range of phosphorus loading rates (L_p) at a fixed dilution rate ($D = 0.01\ day^{-1}$, as in previous models), we find that the system generally has a strong periodic attractor over the major part of the loading gradient. Taking long-term averages over the periodic orbits, we can illustrate some general properties of the system by considering the partitioning of total phosphorus and organic carbon as functions of P loading (Figs. 6.23 and 6.24).

The present model also has a critical loading rate, below which primary production is insufficient to support positive net zooplankton growth (Fig. 6.17). In common with previous models, this critical loading level appears to be below what is normally encountered in lakes. Above this critical loading level, the zooplankton fraction of total P increases steeply to a maximum, and then decreases at high loading rates. The establishment of a grazer population is followed by a sharp increase in the bacterial fraction of total P, with a gradual displacement of total P from the bacterial to the algal compartment with further increase in the phosphorus loading.

Comparing the average partitioning of total P (Fig. 6.17) with the corresponding distribution of total organic C (Fig. 6.18) reveals a striking difference: while bacteria and grazers constitute a major fraction of total P, their share of total organic C is always less than 25%. The model predicts that algal biomass should constitute the largest carbon pool at all loading rates, with the nonliving pools of particulate detritus and dissolved organic carbon ranking second.

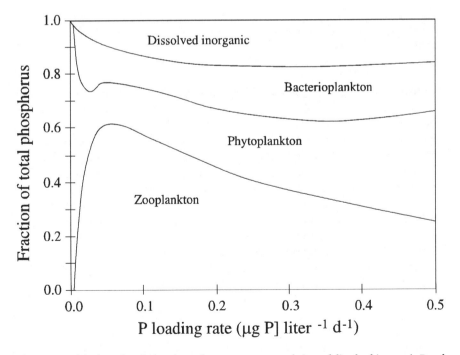

Fig. 6.17. Partitioning of total phosphorus between average pool sizes of dissolved inorganic P and particulate P contained in algae, bacteria, and grazers in a gradient of phosphorus loading rates

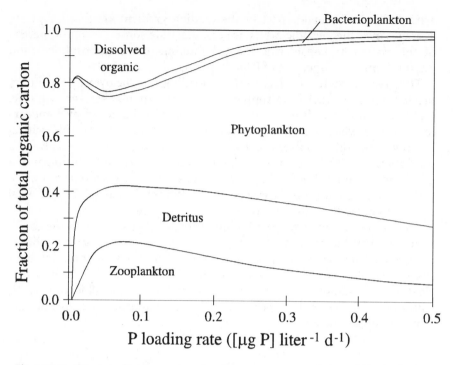

Fig. 6.18. Partitioning of total organic carbon between average pool sizes of dissolved organic C and particulate C contained in algae, bacteria, detritus, and grazers in a gradient of phosphorus loading rates

Eutrophication as a Period-Doubling Cascade. The representations of long-term averages in Figs. 6.17 and 6.18 hide important dynamic aspects of the present model. While models presented elsewhere in this work generally have simple attractors that are either stable equilibrium points or limit cycles, the present model introduces qualitatively new modes of dynamic behavior. For certain loading rates, such as in Fig. 6.19, the system displays persistent oscillations which are not closed periodic orbits: instead, consecutive peaks of food carbon vary by at least a factor of 3, while different minima in zooplankton biomass vary by a factor of 2. The system is apparently never returned to exactly the same state after each completed cycle, leading to a nonintersecting, aperiodic trajectory in the phase space. In the projection of Fig. 6.19, trajectories appear to come arbitrarily close at certain times while spreading out over a surface in other parts of the cycle.

In the language of nonlinear dynamics, the phase portrait in Fig. 6.19 has the appearance of a strange attractor, that is, an attractor which is not a simple geometric structure like a closed curve or a surface, and whose dimensionality therefore is noninteger, or fractal. Strange attractors are usually, but not always (Tufillaro et al. 1992), associated with systems exhibiting

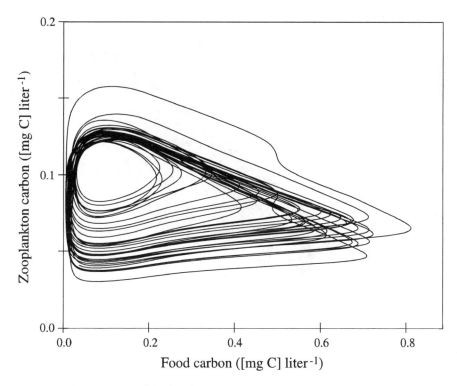

Fig. 6.19. Phase portrait of food carbon (as the sum of algae, bacteria and detritus) and zooplankton biomass at a loading rate of 0.1 (μg P) l^{-1} day^{-1}

deterministic chaos. Chaotic systems are characterized by a sensitivity to initial conditions such that trajectories originating from initial values arbitrarily close to each other will diverge exponentially in time (that is, they have at least one positive Lyapunov exponent; Thompson and Stewart 1986).

The transition to chaos is usually found to follow a characteristic sequence of bifurcations, often called a period-doubling cascade (Tufillaro et al. 1992), where a closed periodic orbit is expanded into a geometric progression of 2-, 4-, 8-cycles, etc. (Fig. 6.21) until a point where cycles of different period are mixed in a chaotic motion. If we use the peaks in zooplankton biomass as delimiters for individual cycles (after transients have died out), we can construct a bifurcation diagram by plotting successive cycle averages of a state variable against the bifurcation parameter, which in our case will be phosphorus loading rate. In such a diagram, a simple periodic orbit, or limit cycle, will be represented as a single point, a general n-cycle by n points, and an aperiodic orbit by a dense set of points forming a solid vertical line. Figure 6.20 shows that the period doubling cascade starts from simple limit cycle at a critical loading rate $L_p \approx 0.05$ (μg P) l^{-1} day^{-1} and develops into aperiodic motion when the loading rate exceeds ≈ 0.066 (μg P) l^{-1} day^{-1}. Above this

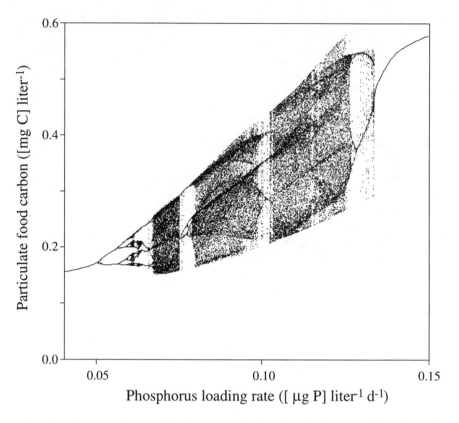

Fig. 6.20. Bifurcation diagram showing average food carbon (as the sum of algae, bacteria, and detritus) for individual zooplankton biomass cycles as function of phosphorus loading rate

loading level, intervals of chaotic motion are found interspersed with windows of periodic orbits that give rise to new period doubling sequences. Aperiodic orbits persist until the loading rate exceeds a critical upper limit around 0.133 (μg P) l^{-1} day^{-1}, where suddenly all orbits collapse into a single limit cycle.

A more detailed view of the strange attractor is given by Fig. 6.22, which shows the time course of the distribution of particulate carbon in an aperiodic orbit generated at a loading rate of 0.11 (μg P) l^{-1} day^{-1}. The trajectory in Fig. 6.22 shows an overall pattern of prey-predator oscillations between grazers and the food compartments (algae, detritus and bacteria). While the lengths of individual periods fluctuate irregularly between 22 and 36 days, the general course of events in any individual cycle seems to follow a regular pattern.

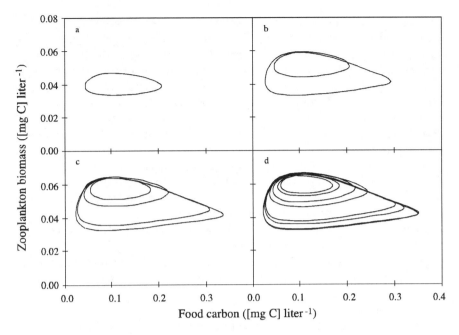

Fig. 6.21a-d. First four steps of the period-doubling cascade in Fig. 6.20. **a** 1-cycle at 0.46 (μg P) l^{-1} day^{-1}. **b** 2-cycle at 0.55 (μg P) l^{-1} day^{-1}. **c** 4-cycle at 0.58 (μg P) l^{-1} day^{-1}. **d** 8-cycle at 0.59 (μg P) l^{-1} day^{-1}

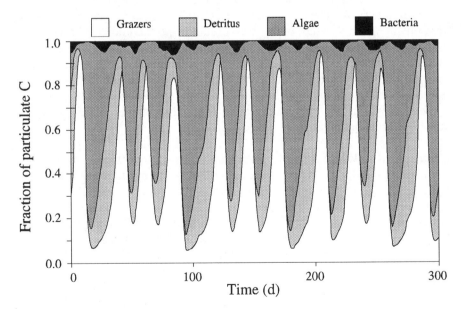

Fig. 6.22. Relative composition of particulate carbon in a trajectory belonging to the aperiodic attractor at a loading rate of 0.11 (μg P) l^{-1} day^{-1}

After the peak in relative zooplankton C the system is overgrazed: food C is inadequate to support grazer growth, thus making the grazer population decline. High grazing loss rates are reflected by high algal growth rates and, correspondingly, high algal P:C ratios. P-rich food means efficient food carbon utilization and a high P : C ratio in zooplankton release products. The high P:C supply ratio gives a competitive disadvantage of bacteria relative to algae, and gives a net loss rate from the detritus pool. Both detritus and bacteria therefore decline while the relative abundance of algal carbon increases.

Grazing loss rates diminish in proportion to the grazer population decline, thus decreasing algal growth rates and, correspondingly, algal P:C ratio. The decrease in food P content leads to increased grazer carbon egestion and to a net accumulation of detritus carbon. The low P:C ratio in zooplankton release products is to the advantage of bacteria, which can compete more efficiently for phosphorus when the supply rate of dissolved organic carbon is high. On a diet composed mainly of detrital C and bacterial P, the grazers are again able to attain positive net growth and build up a new biomass peak, thus completing the cycle.

The dependency on nonalgal food in the accumulation phase preceding a peak in grazer biomass has the effect of decoupling secondary production from the instantaneous primary production. Secondary production is instead supported, at least partly, by primary production that has been delayed though the detrital pathways of the carbon cycle. Since time delays are generally found to have a destabilizing effect on dynamical systems (MacDonald 1978), the present model potentially contains two sources of dynamical instability; that of the prey-predator interaction itself and that of the time-delayed coupling between primary and secondary production. Such interactions between two or more oscillating processes with incommensurable frequencies (that is, the ratio between the two frequencies is not a rational number) have long been recognized as a common source of chaotic behavior in dynamic systems (e.g., Guckenheimer and Holmes 1983; Thompson and Stewart 1986).

Chaos - and So What? A strictly formal assertion of deterministic chaos in a dynamical system requires an extended set of computational measures, such as computing the fractal dimension of the attractor (e.g., Grassberger 1990) and the Lyapunov exponents of the flow (e.g., Souza-Machado et al. 1990). Although such tests have not been performed, the appearance of Figs. 6.25-6.28 strongly suggests that the present system has chaotic properties, and that these properties might be caused by time-delayed prey-predator interactions that are the results of nonlinear storage effects in the particulate and dissolved detrital compartments. Some indication of the sensitivity to initial conditions in the present model is given by Fig. 6.23, which shows two trajectories resulting from adding 1% random noise to the same initial conditions located on the aperiodic attractor.

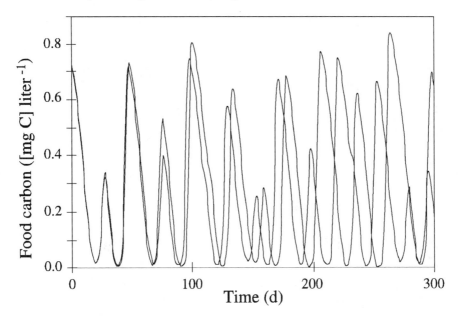

Fig. 6.23. Time course of food carbon (algae, bacteria, and detritus) in two simulation runs differing only by the addition of 1% uniformly distributed random noise to the initial conditions. P loading rate = 0.11 $(\mu g\ P)\ l^{-1}\ day^{-1}$

The trajectories in Fig. 6.23 follow each other quite closely for the first 80 days or so, and then diverge increasingly with time. After about 150 days of simulation time the trajectories are entirely out of phase, thus making the system essentially unpredictable above this time scale. It should be noticed that the divergence of trajectories in Fig. 6.23 would result from any initial condition on the aperiodic attractor. It is therefore qualitatively different from the divergence shown in Fig. 5.9, resulting from choosing a particular pair of initial conditions close to the boundary of the focal attraction basin. While a 1% measurement uncertainty is far better than what it is usually possible to obtain when observing natural communities, even this level of observational variance seems sufficient to make the system unpredictable for time scales longer than 4-6 months.

Figure 6.20 shows that the period-doubling cascade leading to aperiodic cycles is restricted to a limited range of the loading gradient and that the importance of the phenomenon might therefore be somewhat exaggerated. To obtain an impression of the magnitude of aperiodic fluctuations relative to the long-term averages of the state variables, we can compute both means and ranges of state variables for time periods much longer than the average cycle period. Figures 6.15 and 6.16 show that the biomass fluctuations generated by the chaotic attractor are indeed minor compared to the changes in long-term averages caused by the loading conditions.

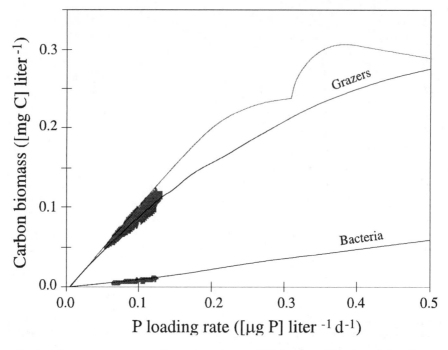

Fig. 6.24. Long-term averages of bacteria and grazer biomasses as function of phosphorus loading. *Shaded areas* Ranges of cycle averages in the chaotic attractor; *dashed curve* corresponding grazer biomass - loading relationship from the model in Chapter 5

The minimal P-cycling model, studied in Chapter 5, predicted that zooplankton biomass should increase in proportion to the P load at low P supply rates and then level off to an asymptotic value at high P loading. Figure 6.24 shows that the inclusion of C-cycling preserves this general pattern, but with the average zooplankton biomass level being some 20-25% lower, and without the characteristic jump discontinuity at the transition to limit cycling behavior. If we consider the sum of zooplankton and bacterial biomasses, the level of total heterotrophic biomass becomes even closer to the predicted relationship between consumer biomass and phosphorus loading from Chapter 5.

Instead of a sharp transition from dominance by a stable equilibrium point to a periodic orbit at a critical bifurcation loading level, the present model predicts a sequence of period-doubling bifurcations leading to a loading interval dominated by aperiodic orbits which then coalesce into a single periodic orbit. This route to prey-predator oscillations creates a smoother relationship between average algal biomass and P loading than in the minimal model of Chapter 5, with the two curves intersecting at a loading rate around 0.4 (μg P) l^{-1} day^{-1} (Fig. 6.25). This means that the loading criterion introduced in Chapter 5, which was based on the bifurcation

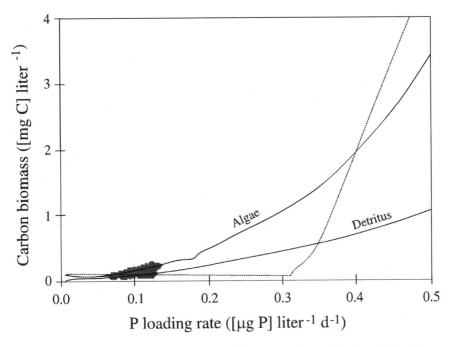

Fig. 6.25. Long-term averages of algal biomass and detritus carbon as function of phosphorus loading. *Shaded areas* Ranges of cycle averages in the chaotic attractor; *dashed curve* corresponding relationship between algal biomass and phosphorus loading from the model in Chapter 5

loading level, cannot be directly translated to the present model. Since the algal biomass–loading relationship is monotonic increasing, we can still set a critical algal biomass such that biomanipulations leading to long-term averages above this level are classified as unsuccessful. If we choose a critical level around 1 (mg C) l^{-1}, a loading diagram similar to Fig. 5.17 can be constructed from the present model.

The possibility of deterministic chaos in ecological systems has led several authors (e.g., Schaffer and Kot 1985) to propose that much of the apparently stochastic fluctuations in natural populations might actually be caused by the internal dynamics of the system, and not by external perturbations. Comparing with Fig. 5.18 and 5.19, it seems evident that chaotic dynamics of a kind exhibited by the present model can explain only a minor part of the variance in observed phyto- and zooplankton biomasses. Since the fluctuations of deterministic chaos seem to be subordinate to the constraints of nutrient loading, the presence of chaotic attractors should have only minor influence on the predictability of long-term biomass relationships. On the other hand, Fig. 6.23 suggests that, under certain conditions, the possibility of chaotic dynamics can still put strong constraints on the predictability of succession and seasonal development in plankton communities.

6.5 Summary and Conclusions

The models presented in this Chapter represent the first few steps towards extending the formalism of pelagic nutrient cycling, as developed in Chapter 2, to real food webs. All models focus on aspects of resource competition among the organisms constituting the food source for secondary producers and thus forming the base of the food web. The main issue has been to investigate the mechanisms by which grazing can promote or impede coexistence among prey species, and how grazing effects on plankton community structure can be influenced by the external nutrient loading conditions.

The basic assumption behind all studies of interspecific competition is the existence of evolutionary tradeoffs between different competitive traits, thus excluding the possibility of a "superspecies" which is competitively superior under all conditions. Different models in this Chapter have focused on the effects of competitive tradeoffs with respect to: nutrient uptake and growth capacity (Section 6.1), predator defence and competitive ability for limiting nutrients (Section 6.2), uptake and utilization of different limiting nutrients (Section 6.3).

In contrast to the model of Chapter 5, which was sufficiently simple to allow formal stability analysis of all equilibrium points, all the models considered in this Chapter have too many state variables (from six to nine) to make such analysis feasible. The analysis of stationary states has therefore necessarily been more heuristic than in Chapter 5, relying heavily on numerical techniques, such as exhaustive simulations from random initial conditions in the search for stationary states and reciprocal invasion experiments to identify conditions for coexistence and competitive exclusion.

The simulation experiments in Section 6.1 show that when the stationary state of the system is in a periodic orbit, it becomes possible for up to three phytoplankton species to coexist on a single limiting nutrient. Since the coexistence is caused by the species exploiting different phases of the prey-predator cycle, it is argued that this situation of predator-mediated coexistence does not represent a strict violation of the competitive exclusion principle as stated by Hardin (1961). The models presented in Sections 6.2 and 6.3 demonstrate two mechanisms that seem to counteract such predator-mediated coexistence.

In the model of Section 6.2, where grazers are assumed to have different feeding efficiencies on different prey species, no predator-mediated coexistence is possible, independently of whether the system is attracted to a stable equilibrium or a periodic orbit. The model is therefore unable to represent the often observed coexistence among ungrazed "canopy" species and edible "undergrowth" species, sensu Sommer et al. (1986). If the large, inedible species are able to migrate vertically (like large dino-flagellates and colonial cyanobacteria), coexistence could result from "canopy" and "undergrowth" species exploiting spatially separated resources. Such a scenario would be impossible to represent in spatially homogeneous environment implicitly assumed in the present models.

In one of the most celebrated results of theoretical plankton ecology, it is inferred that fluctuating resource ratios could support the coexistence of an arbitrary number of species (Tilman 1982) on a small number of resources. In Section 6.3 it is shown that for N and P as the two potentially limiting nutrients, differential nutrient recycling by grazers will lead to diverging resource ratios and thus to competitive exclusion, even when the species are able to coexist in the absence of grazers.

The emergence of chaotic dynamics in the carbon cycling model of Section 6.4 introduces a qualitatively new source of endogenous fluctuations in resource supply rates, which might conceivably give rise to a wider set of temporal niches for different phytoplankton species, although the question of whether prey-predator dynamics on aperiodic attractors really can support more prey species than classical periodic orbits has not yet been resolved.

Classical models of food-chain productivity (Hairston et al. 1960; Oksanen et al. 1981) require that competition and predation can be considered as mutually exclusive; that is, a single trophic level is controlled either from the trophic level below, by competition for limiting resources, or from above, by predators on the next trophic level. Since there are no carnivore levels in the models investigated in this Chapter, the zooplankton can obviously not be controlled by predation. Still, phytoplankton does not appear to be strictly controlled by predation, since competitive exclusion leading to species replacements in the algal community occurs as function of the nutrient loading conditions. The model considered in Section 6.2 illustrates that the ability of a selective grazer to tilt the balance between competing phytoplankton species depends on both the exact nature of the competitive tradeoffs between the prey species and the nutrient supply to the system.

A key feature of all the models presented in this work is that the availability of essential elements can potentially limit secondary production. Both simulations and observations (Hessen et al. 1992) indicate that a significant fraction of total nutrients can be contained in zooplankton biomass, and thus be unavailable to phytoplankton. The considerations of stoichiometric constraints on grazer production and abundance suggest that, in contrast to the view advocated by Harris (1986), heavy grazing and high turnover rates of algal biomass do not necessarily preclude nutrient limitation and, consequently, resource competition.

The bold assertion of Sommer (1989), cited in the introduction to this Chapter, is only partly supported by the models presented here. Although the interaction between predators and their prey can, under certain assumptions, promote the coexistence of several algal species on a single limiting resource, this conclusion does not appear to be structurally stable under the influences of differential grazing or differential nutrient recycling. This makes it likely that fluctuations in resource supply rates generated by simple prey-predator cycles are insufficient to maintain the observed diversity of phytoplankton communities. In other words, the original theory put forward by Hutchinson (1961), that physical disturbances prevent plankton communities from reaching an equilibrium situation leading to competitive exclusion, still remains a strong candidate for explaining the paradox of the plankton.

6.6 Symbols, Definitions, and Units

Symbol	Definition	Unit
α'	Specific nutrient uptake affinity	$l\,(mg\,C)^{-1}\,day^{-1}$
C	Food carbon biomass	$(mg\,C)\,l^{-1}$
C_a	Algal carbon biomass	$(mg\,C)\,l^{-1}$
C_b	Bacterial carbon biomass	$(mg\,C)\,l^{-1}$
C_d	Particulate detritus carbon	$(mg\,C)\,l^{-1}$
C'	Incipient limiting food concentration	$(mg\,C)\,l^{-1}$
δ	Specific grazer mortality rate	day^{-1}
D	Dilution rate	day^{-1}
ε	Grazer food carbon assimilation efficiency	–
ε_b	Bacterial carbon assimilation efficiency	–
f_e	Fraction extracellular release of total primary production	–
f_p	Particulate fraction of grazer organic C release	–
F	Specific grazer clearance rate	$l\,(mg\,C)^{-1}\,day^{-1}$
F_i	Specific clearance rate on prey species i	$l\,(mg\,C)^{-1}\,day^{-1}$
ϕ	Grazer feeding selectivity index	–
ϕ^*	Critical feeding selectivity for a pair of algal competitors	–
g	Grazer growth rate	day^{-1}
g'	Maximum grazer growth rate on optimally composed food	day^{-1}
I	Specific food ingestion rate	day^{-1}
I'	Maximum specific ingestion rate at saturating food	day^{-1}

Symbol	Definition	Unit
K'	Monod half-saturation parameter for P-limited growth	$(\mu g\ P)\ l^{-1}$
L_N	Volumetric N loading rate	$(\mu g\ N)\ l^{-1}\ day^{-1}$
L_p	Volumetric P loading rate	$(\mu g\ P)\ l^{-1}\ day^{-1}$
m	Specific loss rate due to grazing	day^{-1}
N	Concentration of algal-bound N	$(\mu g\ N)\ l^{-1}$
N_L	Input N concentration	$(\mu g\ N)\ l^{-1}$
μ	Algal growth rate	day^{-1}
μ_a	Algal growth rate	day^{-1}
μ_b	Bacterial growth rate	day^{-1}
$\mu_{C,b}$	C-limited bacterial growth rate	day^{-1}
$\mu_{P,b}$	P-limited bacterial growth rate	day^{-1}
μ'	Asymptotic algal growth rate in the Droop model	day^{-1}
μ''	Maximal algal growth rate	day^{-1}
θ	Grazer P content	$(\mu g\ P)\ (mg\ C)^{-1}$
θ_N	Grazer N content	$(\mu g\ N)\ (mg\ C)^{-1}$
θ_P	Grazer P content	$(\mu g\ P)\ (mg\ C)^{-1}$
θ_R	Grazer element R content	$(\mu g\ R)\ (mg\ C)^{-1}$
Q	Algal P content	$(\mu g\ P)\ (mg\ C)^{-1}$
Q'	Minimum algal P content (subsistence quota)	$(\mu g\ P)\ (mg\ C)^{-1}$
Q''	Maximum algal P content	$(\mu g\ P)\ (mg\ C)^{-1}$
Q_P	Food P content	$(\mu g\ P)\ (mg\ C)^{-1}$
Q_N	Food N content	$(\mu g\ N)\ (mg\ C)^{-1}$
Q_R	Food element R content	$(\mu g\ R)\ (mg\ C)^{-1}$
P	Concentration of algal-bound P	$(\mu g\ P)\ l^{-1}$
P_a	Concentration of algal-bound P	$(\mu g\ P)\ l^{-1}$
P_b	Concentration of bacterial-bound P	$(\mu g\ P)\ l^{-1}$
P_L	Input P concentration	$(\mu g\ P)\ l^{-1}$

Symbol	Definition	Unit
ρ	Specific grazer P release rate	$(\mu g\,P)\,(mg\,C)^{-1}\,day^{-1}$
ρ_N	Specific grazer N release rate	$(\mu g\,N)\,(mg\,C)\,day^{-1}$
ρ_P	Specific grazer P release rate	$(\mu g\,P)\,(mg\,C)\,day^{-1}$
ρ_C	Specific grazer organic C release rate	day^{-1}
ρ_R	Specific grazer element R release rate	$(\mu g\,R)\,(mg\,C)\,day^{-1}$
r	Specific grazer respiration rate	day^{-1}
r_b	Specific bacterial maintenance respiration rate	day^{-1}
R_L	Input concentration of element R	$(\mu g\,R)\,l^{-1}$
σ	Algal sinking loss rate	day^{-1}
S	Dissolved inorganic P concentration	$(\mu g\,P)\,l^{-1}$
S_P	Dissolved inorganic P concentration	$(\mu g\,P)\,l^{-1}$
S'	Threshold DIP concentration for positive net uptake	$(\mu g\,P)\,l^{-1}$
S_N	Dissolved inorganic N concentration	$(\mu g\,N)\,l^{-1}$
S_C	Dissolved organic C concentration	$(mg\,C)\,l^{-1}$
S_R	Concentration of element R in dissolved inorganic form	$(\mu g\,R)\,l^{-1}$
u_b	Specific C uptake rate in bacteria	day^{-1}
u''_b	Maximum bacterial C uptake rate	day^{-1}
v	Specific P uptake rate	$(\mu g\,P)\,(mg\,C)^{-1}\,day^{-1}$
Z	Grazer carbon biomass	$(mg\,C)\,l^{-1}$

7 Grazers as Sources and Sinks for Nutrients: Conclusions, Limitations, and Speculations

> *If further progress should be possible, then more complex models are needed. ... Also, the trophic-dynamic interrelationships in the sense of Lindeman (1942) requires much more sophisticated analyses.*
>
> R. A. Vollenweider (1976).

The trophic-dynamic school in ecology, pioneered by Lindeman (1942), has typically used mass or energy equivalents for the quantitative description of interactions between trophic levels. Progress in the direction envisioned by Vollenweider (1976), of coupling input-output models with a trophic-dynamic view of ecosystems, would require also the consideration of flows of limiting nutrients among trophic levels. The flow networks of energy and essential elements cannot be simple mirror images of each other, since, for example, energy flows unidirectionally from autotrophs to herbivores, while nutrients flow bidirectionally between the same two trophic levels. As pointed out by Reiners (1986), the energetic and stoichiometric views of ecosystem organization, as expressed by e.g., Odum (1957) and Redfield (1958), must therefore be seen as complementary and irreducible parts of models describing trophic relationships in nutrient-limited ecosystems.

Herbivorous zooplankton hold a key position in pelagic food webs both by consuming primary producers, and by supplying regenerated nutrients supporting primary production. Exploring this important dualism in the functional role of zooplankton, as both sources and sinks for nutrients, has been a central theme throughout this work. This analysis has been facilitated by applying the homeostatic view of zooplankton stoichiometry supported by Sterner (1990) and Andersen and Hessen (1991): that the element composition of zooplankton can be considered constant and species-specific, at least compared to the wide variability of elemental ratios found in the unicellular organisms constituting their food. In Chapter 2 of this work, it is shown that this hypothesis leads to a general representation of nutrient cycling and nutrient partitioning in plankton communities, that can be regarded as a direct extension of the one-compartment phosphorus loading model of Vollenweider (1976).

Whether the zooplankton acts as a source or a sink for nutrients in a specific situation depends on both the temporal scale and on the frame of reference. Nutrients contained in an individual zooplankter will have a high likelihood of being lost from the pelagic zone upon death of the organism. Hence, on a sufficiently long-termscale, zooplankton will always represent a net sink in the total nutrient mass balance of a lake. The actual quantitative importance of zooplankton as a sink in the total nutrient budget will depend on both the amount of nutrient located in zooplankton and on the dominating fate of individual organisms.

All the models considered in this work, as well as a recent regional survey (Hessen et al. 1992), suggest that the fraction of total nutrients located in zooplankton biomass will increase to a maximum at some intermediate nutrient loading rate, and then decrease with further enrichment. Since loss rates from natural zooplankton mortality appear to be high compared to observed net loss rates of total phosphorus, it is predicted that phosphorus retention should also follow a unimodal pattern with a maximum of zooplankton-mediated nutrient losses at some intermediate loading level. Such a mechanism might possibly explain a substantial part of the order-of-magnitude variability in phosphorus retention that is found among lakes with the same water renewal rate (Fig. 2.2).

In the presence of zooplankton predators, one would expect a reduced importance of zooplankton mortality in total nutrient losses, since nutrients in dead zooplankters should then, to a larger extent, either be recycled or incorporated into predator biomass, instead of being lost from the pelagic zone. Changes in zooplankton-mediated retention offer an explanation to the observed reductions in total phosphorus after successful biomanipulation treatments (Wright and Shapiro 1984; Reinertsen et al. 1989; Sanni and Wærvågen 1990). The decreasing importance of zooplankton mortality as a retention process at very high nutrient loadings, predicted by the present model (Fig. 5.16), might explain the contrasting observations of Jeppesen et al. (1990), who found that biomanipulation did not reduce total phosphorus in several hypereutrophic Danish lakes.

When seen from the perspective of the whole phytoplankton community, grazers will be both a sink in the sense that they constitute a loss process in algal mass budget, and a source in the sense that they supply the algae with regenerated nutrients. On a sufficiently long-term scale, zooplankton will nevertheless be a net sink of nutrients from the phytoplankton as a whole. Comparison of the range of phosphorus subsistence quotas and storage capacities in phytoplankton (Figs. 3.3 and 3.4) with typical zooplankton element ratios (Andersen and Hessen 1991) indicates that P-limited food algae should often be phosphorus-deficient relative to the requirements for balanced grazer growth. If zooplankton were able to reclaim all metabolic wastes when element composition of the food particles is inadequate for balanced growth, grazers could in certain situations also constitute a gross sink for phosphorus (i.e. no source at all). The analysis of P-release rates in Section 4.5 suggests that the phosphorus economy of *Daphnia* is less efficient than this extreme, and that we thus should expect some P to be released from grazers, at least as long as the food P content remains nonzero.

In a simple two-level food chain, such as the minimal P-cycling model described in Chapter 5, we should expect an efficient channeling of nutrient input from primary to secondary producers. In an eutrophication gradient of increasing nutrient loadings, this should lead to equilibrium zooplankton biomass increasing in proportion to primary production, without any corresponding increase in phytoplankton biomass. Since algal P content is intimately related to equilibrium growth rate, we would expect the nutritional value of algae as zooplankton food to increase in a gradient of nutrient loading rates. Zooplankton should therefore recycle an increasing fraction of P contained in ingested food, and thus, function more as a source than as a sink to the phytoplankton with increasing eutrophication.

Stoichiometric considerations of the phosphorus economy of grazers suggest that mineral nutrient limitation of secondary production is possible. The model analysis in Section 5.3 nevertheless shows that an equilibrium situation with grazer growth limited by food quality, but not food quantity, will be dynamically unstable. The existence of such an unstable equilibrium point would imply also the existence of a stable equilibrium point with zero grazer biomass, and thus the possibility of deterministic extinction of the grazer population. Since the likelihood of grazer extinction seems to increase with nutrient loading, the consideration of predator stoichiometry leads to a reinforcement of Rosenzweig´s (1971) "paradox of enrichment".

Grazer extinction is possible when grazers are unable to attain positive net growth, and thus unable to invade a phytoplankton community growing in equilibrium with dilution and sinking losses. This condition defines a persistence boundary in terms of a critical water renewal rate, above which grazer extinction becomes possible. With the set of parameters used in the present models, the majority of proper lakes seem to have average dilution rates well below this critical level. On the other hand, parameter sensitivity analysis of the persistence boundary suggests that we should expect grazer populations to be most likely to be driven to extinction by predatory mortality in lakes with high dilution rates.

Since algal growth rates are constrained below an upper physiological limit, there must also be an upper limit to the secondary production that can be supported by a given algal biomass. Consequently, the equilibrium level of zooplankton biomass in a simple two-level food chain cannot increase indefinitely in proportion to nutrient loading, while maintaining algal biomass compatible with food-limited grazer growth, but must instead approach an asymptotic limit. This means that with increasing P loading, an increasing fraction of total phosphorus will be available for algal growth, and that algae will be able to depart further and further from the equilibrium level if grazer biomass at any time is insufficient to control their growth. As a result, there will be a critical loading level where persistent oscillations and alternating over- and undergrazing of algal biomass becomes an alternative stationary state to the simple equilibrium point with constant biomasses. In a sense, eutrophication can therefore be regarded as a bifurcation parameter determining the qualitative dynamic behavior of the system.

A major point made in this work is that such periodic orbits are not necessarily symmetrical around the equilibrium point. Instead, differences in magnitude and duration between algal and grazer peaks make the long-term average biomasses diverge between the two modes of dynamic behavior. This causes phyto- and zooplankton biomasses to increase super- and sublinearly with nutrient loading, respectively, a pattern that seems to be in accordance with observed phosphorus–biomass relationships (Figs. 5.19 and 5.20). The loading level corresponding to the emergence of a periodic orbit, called the bifurcation level, can be seen as a point of departure from strict grazer control of algal biomass.

Since the maintenance of stable grazer control on algal abundance is the major goal of biomanipulation, the bifurcation level can be used to develop criteria for assessing the likelihood of water quality improvement by bio-manipulation in a given lake. Although the resulting criteria (Figs. 5.16 and 5.17) show clear resemblance to classical loading diagrams, the proposed feasible region for biomanipulation actually encloses all the lakes considered by Vollenweider (1976). Due to the striking lack of loading data from lakes that have been subject to biomanipulation experiments, the prospects for a rigorous verification of the model predictions are so far limited.

Perhaps the most interesting device for further model testing is through the use of artificial food chains and large-volume laboratory ecosystems, where loading and dilution rates can be more easily controlled and measured than in lake-scale experiments. Borgmann et al. (1988) cultured *Daphnia magna* and several species of algae in a large (3.4 m^3) laboratory ecosystem under different phosphorus supply and dilution rates. This system exhibited damped oscillations towards a stable equilibrium for P supply rates ≤ 0.3 (μg P) l^{-1} day^{-1} and dilution rates ≤ 0.037 day^{-1}, persistent oscillations at the highest loading rate tested [0.74 (μg P) l^{-1} day^{-1}], and grazer extinction at the highest dilution rate tested (0.072 day^{-1}). These critical loading and dilution rates correspond surprisingly well with the bifurcation loading level and the persistence boundary defined in the models presented here.

The success of Borgmann et al. (1988) in maintaining stable artificial food chains for up to 20 weeks is probably due their choice of dilution and nutrient supply rates that might seem very small to those familiar with chemostat experiments with plankton algae, but seem nevertheless to be matching typical loading and dilution rates found in lakes. Dilution rates should in particular be given closer consideration in designing artificial food chains in microcosm experiments. The stoichiometric constraints represented by the persistence boundary might, for example, offer an explanation why Elstad (1986) was unable to establish a stable *Daphnia* population in a system where the dilution rate in the algal compartment was 0.05 day^{-1} (which is higher than the critical level for grazer persistence, according to the model in Chap. 5).

The positive relationship between algal growth rate and nutrient supply rate in the simple two-level food chains considered here implies that eutrophication can also be considered a gradient in loss rates from the phytoplankton community. This means that in the stable equilibrium situation, the outcome of algal resource competition and the sequence of species replacements in an eutrophication gradient are directly predictable from the Monod curves for the individual species (Figs. 3.6 and 6.2). When the system has a stationary state with persistent (periodic or aperiodic) oscillations, the zooplankton can change temporally from being a net sink, when the population is increasing, to a net source when it is declining. In a situation with several competing prey populations, a fluctuating grazer biomass can alternate between being a net source and a net sink for nutrients to different prey populations. This creates temporal niches that can be exploited by different algal species, allowing coexistence of up to three species on a single limiting resource (Fig. 6.4).

Such coexistence, mediated by fluctuations in grazer abundance, is apparently only possible when the grazers feed nonselectively on the available prey species: introducing either differential grazing or differential recycling of essential resources leads to the competitive exclusion of all but one species (Sects.6.2 and 6.3). In such a competitive situation, grazers will generally constitute a net source for nutrients to the superior species, and a net sink to the excluded species, irrespective of whether the grazer population is stable or fluctuating.

Since nutrients located in inedible algae will be unavailable to grazers, such species will represent a net sink for nutrients from the herbivore point of view. The accumulation of inedible algae will reduce food-chain transfer efficiency of energy and nutrients, and eventually destroy food-chain integrity; as such, it is often considered a major cause of biomanipulation failure. Since edible algae suffer extra losses in the presence of grazers, food-chain integrity can be maintained if edible algae are competitively superior to inedible algae in the absence of grazers. In other words, food chains can be resistant to invasion by inedible species if strong tradeoffs between predator defence and competitive ability for limiting nutrients can be found. The survey of algal growth and nutrient uptake parameters in Chapter 3 revealed only one such tradeoff; algae classified as resistant to grazing had significantly lower maximal specific growth rates. The model predictions in Figs. 6.9 and 6.10 indicate that a growth rate advantage is insufficient to make food chains invasion-resistant, and that uninvadability to completely inedible algae will only be possible if the invaders have additional handicaps, such as high sinking loss rates or high nutrient efflux rates.

Due to the lack of systematic studies with simultaneous measurement of relevant model parameters, individual parameters have mostly been treated as independent, random variables. The demonstration of evolutionary tradeoffs between different model parameters would impose a correlation structure on the parameter space, which most likely would increase the quality of model predictions by reducing the sensitivity to individual model parameters compared to what is indicated by first-order sensitivity analyses, as in e.g. Figs. 4.21 and 5.15. On the other hand, the failure of such a search to reveal any tradeoffs or correlations between competitive traits could probably be taken as a strong argument against the school advocating resource competition as a major force in structuring plankton communities (e.g. Sommer 1989a).

McCauley and Murdoch (1989) have argued against nutrient supply rates as the determinant of *Daphnia* dynamics, promoting the view that *Daphnia* cycles are solely of demographic nature, caused by the dominance and suppression between strong and weak cohorts. Since their argumentation is mostly indirect through using algal biomass as a measure of nutrient enrichment, it is difficult to compare the nutrient supply rates used in their storage tank experiments with the critical levels predicted by the present model. It should nevertheless be pointed out that the coexistence of two alternative stationary states in the form of a stable equilibrium point and a periodic orbit, as in the present models, makes it possible to have qualitatively different grazer dynamics at the same nutrient supply rate, depending on the initial conditions. Thus, the observation of different patterns of *Daphnia* abundance between different years in the same lake (McCauley and Murdoch 1989) does not necessarily prove the unimportance of nutrient supply rates in determining grazer dynamics.

It nevertheless seems clear that the third type of dynamics discovered by Murdoch and McCauley (1985), *Daphnia* oscillations without any phase-lagged cycles in phytoplankton abundance, cannot be given an adequate model representation without taking population structure into consideration. Although a structured *Daphnia* population model would seem a logical consequence of the models of individual growth, reproduction, and mortality in Chapter 4, this step has not been made in the present work. While the formal mathematical framework for constructing such models is well established (Sinko and Streifer 1967; Metz and Diekman 1986), much of the necessary input data are still lacking due to the inadequacies of the transfer culture methods commonly employed in life-table studies of food-limited *Daphnia* growth. In Section 4.6, it is argued that the interactions between the brood cycle and the feeding cycle in such experiments create maternal effects that can effectively mask any relationship between recent feeding history and juvenile growth and survival under food limitation. Progress in this direction will thus require the use of more sophisticated culture techniques to avoid the "feast and famine" cycles characterizing transfer culture methods.

While the representation of fast-growing animals, like cladocerans and rotifers, by unstructured population models might be defendable under certain circumstances, this approximation becomes progressively harder to justify for more long-lived animals, such as carnivorous zooplankton and fish. Since both vertebrate and invertebrate predators are generally size-selective, an extension of the present models to include higher trophic levels would probably require all consumer populations to be represented by structured population models.

The representation of perennial organisms like fish would probably also require a closer consideration of seasonal cycles, which have been completely neglected in the stratified summer situation implicitly assumed in the present models. A realistic representation of seasonal cycles in lakes would then again probably also require consideration of the major physical processes controlling stratification and vertical transport. In principle, such a vertically structured model could also be used to model the evolutionary games involved in vertical migration (e.g. Mangel and Clark 1988), although the computational costs of such models would probably be very high. Some would even argue that the whole three-dimensional flow field should be represented in a realistic lake model, since lateral diffusion and spatial heterogeneity are thought to deeply influence the stability properties of interacting populations (e.g. Powell and Richerson 1985).

While physically and demographically structured models (e.g. Slagstad 1981, 1982) are definitely superior in terms of realism and generality, it is not at all certain whether the added amount of dynamic complexity and parameter uncertainty has a proportional reward in terms of predictive power and biological insight compared to simple, unstructured models. Perhaps a better strategy for future research would be to use minimal models of the kind presented here to design critical experiments and identify sensitive parameters, and then proceed to more complex, structured models when the potential of minimal models has been definitively exhausted.

References

Ahlgren G (1977) Growth of *Oscillatoria agardhii* in chemostat culture. 1. Nitrogen and phosphorus requirements. Oikos 29: 209-224

Ahlgren G (1978) Growth of *Oscillatoria agardhii* in chemostat culture. 2. Dependence of growth constants on temperature. Mitt Int Ver Limnol 21: 88-102

Ahlgren G (1985) Growth of *Microcystis wesenbergii* in batch and chemostat cultures. Verh Int Ver Limnol 22: 2813-2820

Ahlgren G (1987) Temperature functions in biology and their application to algal growth constants. Oikos 49: 177-190

Aksnes D L Egge J K (1991) A theoretical model for nutrient uptake in phytoplankton. Mar Ecol Prog Ser 70: 65-72

Allan J D (1976) Life history patterns in zooplankton. Am Nat 110: 165-180

Allan J D Goulden C E (1980) Some aspects of reproductive variation among freshwater zooplankton. Am Soc Limnol Oceanogr Spec Symp 3: 388-411

Andersen O K Goldman J C Caron D A Dennett M R (1986) Nutrient cycling in a microflagellate food chain: III. Phosphorus dynamics. Mar Ecol Prog Ser 31: 47-55

Andersen T Hessen D O (1991) Carbon, nitrogen, and phosphorus content of freshwater zooplankton. Limnol Oceanogr 36: 807-814

Armstrong R A McGehee R (1980) Competitive exclusion. Am Nat 115: 151-170

Arneodo A Coullet P Peyraud J Tresser C (1982) Strange attractors in Volterra equations for species in competition. J Math Biol 14: 153-157

Arnold D E (1971) Ingestion, assimilation, survival and reproduction by *Daphnia pulex* fed seven species of blue-green algae. Limnol Oceanogr 16: 906-921

Azam F Fenchel T Field J G Gray J S Meyer-Reil L A Thingstad T F (1983) The ecological role of water-column microbes in the sea. Mar Ecol Prog Ser 10: 257-263

Bader F G (1982) Kinetics of double-substrate limited growth. In: Bazin M (ed) Microbial population dynamics. CRC Press, Boca Raton, pp 1-32

Baines S B Pace M L (1991) The production of dissolved organic matter by phytoplankton and its importance to bacteria: patterns across marine and freshwater systems. Limnol Oceanogr 36: 1078-1090

Banse K (1982) Cell volumes, maximal growth rates of unicellular algae and ciliates, and the role of ciliates in the marine pelagial. Limnol Oceanogr 27: 1059-1071

Banse K (1990) New views on the degradation and disposition of organic particles as collected by sediment traps in the open sea. Deep-Sea Res 37: 1177-1195

Banta A M (1939) Studies of the physiology, genetics and evolution of some Cladocera. Carnegie Inst Wash Publ 513: 1-285

Beklemishev C W (1962) Superfluous feeding of marine herbivorous zooplankton. Rapp P V Reun Cons Perm Int Explor Mer 153: 108-113

Benndorf J (1987) Food web manipulation without nutrient control: A useful strategy in lake restoration? Schweiz Z Hydrol 49: 237-248

Benndorf J Kneschke H Kossatz K Penz E (1984) Manipulation of the pelagic food web by stocking with predacious fishes. Int Rev Gesamten Hydrobiol 69: 407-428

Berryman A A Millstein J A (1989) Are ecological systems chaotic - and if not, why not? TREE 4: 26-28

Bird D F Kalff J (1986) Bacterial grazing by planktonic lake algae. Science 231: 493-495

Bjørnsen P K (1986) Bacterioplankton growth yield in continuous plankton cultures. Mar Ecol Prog Ser 30: 191-196

Blueweiss L Fox H Kudzma V Nakashima D Peters R H Sams S (1978) Relationships between body size and some life history parameters. Oecologia 37: 257-272

Bohrer R N Lampert W (1988) Simultaneous measurement of the effect of food concentration on assimilation and respiration in Daphnia magna Straus. Funct Ecol 2: 463-471

Borgmann U Millard E S Charlton C C (1988) Dynamics of a stable, large volume, laboratory ecosystem containing Daphnia and phytoplankton. J Plankton Res 10: 691-713

Boström B Jansson M Forsberg C (1982) Phosphorus release from lake sediments. Arch Hydrobiol Beih 18: 5-59

Bottrell H H (1975) Generation time, length of life, instar duration and frequency of molting, and their relationship to temperature in eight species of Cladocera from the River Thames, Reading. Oecologia 19: 129-140

Brabrand Å Faafeng B Nilssen J P (1990) Relative importance of phosphorus supply to phytoplankton production: fish excretion versus external loading. Can J Fish Aquat Sci 47: 364-372

Bratbak G Thingstad T F (1985) Phytoplankton-bacteria interactions: an apparent paradox? Analysis of a model system with both competition and commensalism. Mar Ecol Prog Ser 25: 23-30

Brekke O (1987) Phosphorus limited growth and phosphate uptake of freshwater algae: Chemostat studies of Rhodomonas lacustris (Cryptophyceae) and competition studies in a eutrophic lake [in norwegian]. Thesis. University of Trondheim, Trondheim

Brooks J L Dodson S I (1965) Predation, body size and composition of plankton. Science 150: 28-35

Brown E J Button D K (1979) Phosphate limited growth kinetics of Selenastrum capricornutum (Chlorophyceae). J Phycol 15: 305-311

Brown E J Harris R F Koonce J F (1978) Kinetics of phosphate uptake by aquatic microorganisms: deviations from a simple Michaelis-Menten equation. Limnol Oceanogr 23: 26-34

Buikema A L (1972) Oxygen consumption of the cladoceran Daphnia pulex, as a function of body size, light, and light acclimation. Comp Biochem Physiol 42: 877-888

Burmaster D E (1979) The continuous culture of phytoplankton: mathematical equivalence among three unsteady state models. Am Nat 113: 123-134

Burns C W Rigler F H (1967) Comparison of filtering rates of Daphnia rosea in lake water and in suspensions of yeast. Limnol Oceanogr 12: 492-502.

Button D K (1978) On the theory of control of microbial growth kinetics by limiting nutrient concentrations. Deep-Sea Res 25: 1163-1177

Canale R P (1970) An analysis of models describing predator-prey interaction. Biotechnol Bioeng 12: 353-378

Caperon J Meyer J (1972) Nitrogen-limited growth of marine phytoplankton. I. Changes in population characteristics with steady-state growth rate. Deep-Sea Res 19: 601-618

Carpenter S R Kitchell J F (1984) Plankton community structure and limnetic primary production. Am Nat 124: 159-172

Cembella A D Antia N J Harrison P J (1984a) The utilization of inorganic and organic phosphorus compounds as nutrients by eucaryotic microalgae: a multidisciplinary perspective: Part 1. CRC Crit Rev Microbiol 10: 317-391

Cembella A D Antia N J Harrison P J (1984b) The utilization of inorganic and organic phosphorus compounds as nutrients by eucaryotic microalgae: a multidisciplinary perspective: Part 2. CRC Crit Rev Microbiol 11: 13-81

Checkley D M Jr. (1980) The egg production of a marine planktonic copepod in relation to its food supply: laboratory studies. Limnol Oceanogr 25: 430-446

Chen M (1974) Kinetics of phosphorus absorption by *Corynebacterium bovis*. Microb Ecol 1: 164-175

Chisholm S W Stross R G (1975) Light/ dark phased cell division in *Euglena gracilis* in PO_4-limited continuous culture. J Phycol 11: 367-373

Cole J J Likens G E Strayer D L (1982) Photosynthetically produced dissolved organic carbon: an important carbon source for planktonic bacteria. Limnol Oceanogr 27: 1080-1090

Condrey R E Fuller D A (1985) Testing equations of ingestion-limited growth. Arch Hydrobiol Beih (Ergebn Limnol) 21: 257-268

Conover R J (1966) Factors affecting the assimilation of organic matter by zooplankton and the question of superfluous feeding. Limnol Oceanogr 11: 346-354

Conover R J (1968) Zooplankton - Life in a nutritionally dilute environment. Am Zool 8: 107-118

Cook J R (1963) Adaptation of growth and division in *Euglena* effected by energy supply. J Protozool 10: 436-444

Corner E D S Davies A G (1971) Plankton as a factor in the nitrogen and phosphorus cycles in the sea. Adv Mar Biol 9: 101-204.

Currie D J Kalff J (1984) A comparison of the abilities of freshwater algae and bacteria to acquire and retain phosphorus. Limnol Oceanogr 29: 298-310

Daniels R E Allan J D (1981) Life table evaluation of chronic exposure to a pesticide. Can J. Fish Aquat Sci 38: 485-494

Dauta A (1982) Conditions de développement du phytoplancton: étude comparative du comportement de huit espèces en culture. I. Détermination des paramètres de croissance en fonction de la lumière et de la température. Ann Limnol 18: 217-262

DeAngelis D L (1990) Dynamics of nutrient cycling and food webs. Chapman & Hall, London, 270 pp

de Bernardi R Peters R H (1987) Why *Daphnia*? Mem Ist Ital Idrobiol 45: 1-9

DeMott W R (1982) Feeding selectivities and relative ingestion rates of *Daphnia* and *Bosmina*. Limnol Oceanogr 27: 518-527

DeMott W R (1985) Relations between filter mesh-size, feeding mode, and particle capture efficiency for cladocerans feeding on ultrafine particles. Arch Hydrobiol Beih (Ergebn Limnol) 21: 125-134

DeMott W R (1989) The role of competition in zooplankton succession. In: Sommer U (ed) Plankton Ecology. Springer, Berlin, Heidelberg, New York, pp 195-252

Den Oude P J Gulati R D (1988) Phosphorus and nitrogen excretion rates of zooplankton from the eutrophic Loosdrecht lakes, with notes on other P sources for phytoplankton requirements. Hydrobiologia 169: 379-390

De Stasio B T Jr (1989) The seed bank of a freshwater crustacean: copepodology for the plant ecologist. Ecology 70: 1377-1389

Dijkstra E W (1976) A discipline of programming. Prentice-Hall, Englewood Cliffs, N. J., 217 pp

Droop M R (1968) Vitamin B_{12} and marine ecology. IV. The kinetics of uptake, growth and inhibition in *Monochrysis lutheri*. J Mar Biol Assoc UK 48: 689-733

Droop M R (1974) The nutrient status of algal cells in continuous culture. J Mar Biol Assoc UK 55: 825-855

Droop M R (1983) 25 years of algal growth kinetics. A personal view. Bot Mar 26: 99-112

Dugdale R C (1967) Nutrient limitation in the sea: dynamics, identification and significance. Limnol Oceanogr 12: 685-695

Dunstan W W Tenore W T (1974) Control of species composition in enriched mass cultures of natural phytoplankton populations. J Appl Ecol 11: 529-536

Durbin E G Durbin A G Smayda T J Verity P J (1983) Food limitation of production by adult *Acartia tonsa* in Narragansett Bay, Rhode Island. Limnol Oceanogr 28: 1199-1213

Edmondson W T Lehman J T (1981) The effects of changes in the nutrient income on the condition of Lake Washington. Limnol Oceanogr 26: 1-29

Elendt B-P (1989) Effects of starvation on growth, reproduction, survival and biochemical composition of *Daphnia magna*. Arch Hydrobiol 116: 415-433

Elrifi I R Turpin D H (1985) Steady-state luxury consumption and the concept of optimum nutrient ratios: a study with phosphate and nitrate limited *Selenastrum minutum* (Chlorophyta). J Phycol 21: 592-602

Elser J J Elser M M MacKay N A Carpenter S R (1988) Zooplankton-mediated transitions between N- and P-limited algal growth. Limnol Oceanogr 33: 1-14

Elstad C A (1986) Population dynamics and nutrient fluxes in an aquatic microcosm. Hydrobiologia 137: 223-237

Eppley R W Sloan P R (1966) Growth rates of marine phytoplankton: correlation with light absorption by cell chlorophyll a. Physiol Plant 19: 47-59

Faafeng B Brettum P Hessen D O (1990) National survey of trophy status of 355 Norwegian lakes [in Norwegian]. Norwegian Institute for Water Research. (Overvåkingsrapport nr 389/90)

Falkowski P G Dubinsky Z Wyman K (1985) Growth-irradiance relationships in phytoplankton. Limnol Oceanogr 30: 311-321

Fogg G E (1966) The extracellular products of algae. Oceanogr Mar Biol Annu Rev 4: 195-212

Frank P W (1952) A laboratory study of intraspecies and interspecies competition in *Daphnia pulicaria* (Forbes) and *Simocephalus vetulus* C. F. Müller. Physiol Zool 25: 178-204

Frank P W Boll C D Kelly R W (1957) Vital statistics of laboratory cultures of *Daphnia pulex* De Geer as related to density. Physiol Zool 30: 287-305

Frauenthal J C (1986) Analysis of age-structure models. In: Hallam T G Levin S A (eds) Mathematical ecology. Springer, Berlin, Heidelberg, New York, pp 117-147

Freedman H I Wolkowicz G S K (1986) Predator-prey systems with group defence: the paradox of enrichment revisited. Bull Math Biol 48: 493-508

Fuchs G W Demmerle S D Canelli E Chen M (1972) Characterization of phosphorus-limited plankton algae. Am Soc Limnol Oceanogr Spec Symp 1: 113-133

Furnas M J (1990) In situ growth rates of marine phytoplankton: approaches to measurement, community and species growth rates. J Plankton Res 12: 1117-1151

Ganf G G Shiel R J (1985) Particle capture by *Daphnia carinata*. Aust J Mar Freshwater Res 36: 371-381

Gard T C Hallam T G (1979) Persistence in food webs: I. Lotka-Volterra food chains. Bull Math Biol 41: 877-891

Geider R J (1987) Light and temperature dependence of the carbon to chlorophyll a ratio in microalgae and cyanobacteria: implications for physiology and growth of phytoplankton. New Phytol 106: 1-34

Geider R J Osborne B A Raven J A (1985) Light effects on growth and photosynthesis of *Phaeodactylum tricornutum*. J Phycol 21: 609-619

Geller W (1975) Die Nahrungsaufnahme von *Daphnia pulex* in Abhängigkeit von der Futterkonzentration, der Temperatur, der Körpergrösse und dem Hungerzustand der Tiere. Arch Hydrobiol Beih 48: 47-107

Gerritsen J Strickler J R (1977) Encounter probabilities and community structure in zooplankton: a mathematical model. J Fish Res Board Can 34: 73-82

Gilpin M E (1972) Enriched predator-prey systems: theoretical stability. Science 177: 902-904

Gilpin M E (1979) Spiral chaos in a predator-prey model. Am Nat 113: 306-308

Glazier D S (1991) Separating the respiration rates of embryos and brooding females of *Daphnia magna*: implications for the cost of brooding and the allometry of metabolic rate. Limnol Oceanogr 36: 354-362

Gliwicz Z M (1986) Predation and the evolution of vertical migration in zooplankton. Nature 320: 746-748

Gliwicz Z M (1990) Why do cladocerans fail to control algal blooms? Hydrobiologia 200/201: 83-97

Gliwicz Z M Lampert W (1990) Food thresholds in *Daphnia* species in the absence and presence of blue-green filaments. Ecology 71: 691-702

Gliwicz Z M Pijanowska J (1989) The role of predation in zooplankton succession. In: Sommer U (ed) Plankton Ecology. Springer, Berlin, Heidelberg, New York, pp 253-296

Goldman J C McCarthy J J (1978) Steady state growth and ammonium uptake of a fast-growing marine diatom. Limnol Oceanogr 23: 695-703

Goldman J C McCarthy J J Peavey D G (1979) Growth rate influence on the chemical composition of phytoplankton in oceanic waters. Nature 279: 210-215

Gons H J (1977) On the light-limited growth of *Scenedesmus protuberans* Fritsch. PhD Thesis. University of Amsterdam, Amsterdam

Gons H J Barug D Goossens H Loogman J G (1978) Growth of *Scenedesmus protuberans* Fritsch in phosphorus-limited continuous cultures. Verh Int Ver Limnol 20: 2308-2313

Goss L B Bunting D L (1980) Temperature effects on zooplankton respiration. Comp Biochem Physiol 66A: 651-658

Gotham I J Rhee G-Y (1981) Comparative kinetics study of phosphate-limited growth and phosphate uptake in phytoplankton in continuous culture. J Phycol 17: 257-265

Goulden C E Hornig L Wilson C (1978) Why do large zooplankton species dominate? Verh Int Ver Limnol 20: 2457-2460

Goulden C E Henry L L Tessier A J (1982) Body size, energy reserves, and competitive ability in three species of Cladocera. Ecology 63: 1780-1789

Goulden C E Henry L Strickler J R (1984) Lipid energy reserves and their role in cladocera.In: Meyers D W Strickler J R (eds) Trophic interactions within aquatic ecosystems. Westview Press, Boulder, pp 167-185

Grassberger P (1990) An optimized box-assisted algorithm for fractal dimensions. Phys Lett A 148: 63-68

Green J (1956) Growth, size, and reproduction in *Daphnia* (Crustacea: Cladocera). Proc Zool Soc Lond 126: 173-204

Grillo J F Gibson J (1979) Regulation of phosphate accumulation in the unicellular cyano-bacterium *Synechococcus*. J Bacteriol 140: 508-517

Grover J P (1989a) Phosphorus-dependent growth kinetics of 11 species of fresh-water algae. Limnol Oceanogr 34: 349-356

Grover J P (1989b) Influence of cell shape and size on algal competitive ability. J Phycol 25: 402-405

Guckenheimer J Holmes P (1983) Nonlinear oscillations, dynamical systems, and bifurcations of vector fields. Springer, Berlin, Heidelberg, New York

Güde H (1985) Influence of phagotrophic processes on the regeneration of nutrients in two-stage continuous culture systems. Microb Ecol 11: 193-204

Gulati R D Lammers E H R R Meijer M-L van Donk E (eds) (1990) Biomanipulation - tool for water management. Kluwer, Dordrecht, 628 pp.

Gurney W S C McCauley E Nisbet R M Murdoch W W (1990) The physiological ecology of *Daphnia*: a dynamic model of growth and reproduction. Ecology 71: 716-732

Hahn G J Meeker W Q (1982) Pitfalls and practical considerations in product life analysis Part I: Basic concepts and dangers of extrapolation. J Qual Technol 14: 144-152

Hairer E Nørsett S P Wanner G (1987) Solving ordinary differential equations. I. Nonstiff problems. Springer, Berlin, Heidelberg, New York, 480 pp

Hairston N C Smith F E Slobodkin L B (1960) Community structure, population control, and competition. Am Nat 94: 421-425

Hall D J (1964) An experimental approach to the dynamics of a natural population of *Daphnia galeata mendotae*. Ecology 45: 94-112

Hall D J Cooper W E Werner E E (1970) An experimental approach to the production dynamics and structure of freshwater animal communities. Limnol Oceanogr 15: 839-928

Hall D J Threlkeld S T Burns C W Crowley P H (1976) The size-efficiency hypothesis and the size structure of zooplankton communities. Annu Rev Ecol Syst 7: 177-208

Hanson J M Peters R H (1984) Empirical prediction of zooplankton and profundal macro-benthos biomass in lakes. Can J Fish Aquat Sci 41:439-455

Hardin G (1960) The competitive exclusion principle. Science 131: 1292-1298

Harris G P (1986) Phytoplankton ecology. Structure, function and fluctuation. Chapman & Hall, London, 384 pp

Harrison P J Davies C O (1979) The use of outdoor phytoplankton continuous cultures to analyze factors influencing species succession. J Exp Mar Biol Ecol 41: 9-23

Healey F P (1985) Interacting effects of light and nutrient limitation on the growth rate of *Synechococcus linearis* (Cyanophyceae). J Phycol 21: 134-146

Healey F P Hendzel L L (1975) Effects of phosphorus defiency of two algae growing in chemostats. J Phycol 11: 303-309

Healey F P Hendzel L L (1979) Indicators of phosphorus and nitrogen deficiency in five algae in culture. J Fish Res Board Can 36: 1364-1369

Healey F P Hendzel L L (1988) Competition for phosphorus between desmids. J Phycol 24: 287-292

Heaney S I Butterwick C (1985) Comparative mechanisms of algal movement in relation to phytoplankton production. In: Rankin M A (ed) Migration: Mechanisms and adaptive significance. University of Texas Press, Austin, pp 114-134

Heaney S I Sommer U (1984) Changes of algal biomass as carbon, cell number and volume, in bottles suspended in Lake Constance. J Plankton Res 6: 239-247

Hebert P D N (1982) Competition in zooplankton communities. Ann Zool Fenn 19: 349-356

Hessen D O (1985a) The relation between bacterial carbon and dissolved humic compounds in oligotrophic lakes. FEMS Microb Ecol 31: 215-223

Hessen D O (1985b) Filtering structures and particle size selection in coexisting cladocera. Oecologia 66: 368-372

Hessen D O (1992) Nutrient element limitation of zooplankton production. Am Nat 104: 799-814

Hessen D O Andersen T (1990) Bacteria as a source of phosphorus for zooplankton. Hydrobiologia 206: 217-223

Hessen D O Andersen T (1992) The algae-grazer interface: feedback mechanisms linked to elemental ratios and nutrient cycling. Arch Hydrobiol Beih (Ergebn Limnol) 35: 111-120

Hessen D O Lyche A (1991) Inter- and intraspecific variations in zooplankton elemental composition. Arch Hydrobiol 121: 343-353

Hessen D O Andersen T Lyche A (1989) Differential grazing and resource utilization of zooplankton in a humic lake. Arch Hydrobiol 114: 321-347

Hessen D O Andersen T Lyche A (1990) Carbon metabolism in a humic lake; pool sizes and cycling through zooplankton. Limnol Oceanogr 35: 84-99

Hessen D O Faafeng B Andersen T (1992) Zooplankton contribution to particulate phosphorus and nitrogen in lakes. J Plankton Res 14: 937-947

Hogeweg P Hesper B (1978) Interactive instruction in population interactions. Comput Biol Med 8: 319-327

Holling C S (1966) The functional response of invertebrate predators to prey density. Mem Entomol Soc Can 48: 1-86

Holm N A Armstrong D E (1981) Role of nutrient limitation and competition in controlling the populations of *Asterionella formosa* and *Microcystis aeruginosa* in semicontinuous culture. Limnol Oceanogr 26: 622-634

Hrbácek J Dvórákova M Korínek V Procházkóva L (1961) Demonstration of the effect of fish stock on the species composition of zooplankton and the intensity of metabolism of the whole plankton association. Verh Int Ver Limnol 14: 192-195

Hsu S B Hubbell S P Waltman P (1977) A mathematical theory for single-nutrient competition in continous cultures of microorganisms. SIAM J Appl Math 32: 366-383

Hutchinson G E (1961) The paradox of the plankton. Am Nat 95: 137-145

Hutchinson G E (1967) A treatise on limnology. vol. II. Wiley, New York

Hutchinson G E (1973) Eutrophication. The scientific background of a contemporary practical problem. Am Sci 61: 269-279

Infante A Litt A H (1985) Differences between two species of *Daphnia* in the use of 10 species of algae in Lake Washington. Limnol Oceanogr 30: 1053-1059

Ingle L Wood T R Banta A M (1937) A study of longevity, growth, reproduction and heart rate in Daphnia longispina as influenced by limitations in quantity of food. J Exp Zool 76: 325-352

Jacobs J (1974) Quantitative measurement of food selection. Oecologia 14: 413-417

Jacoby J M Lynch D D Welch E B Perkins M A (1982) Internal phosphorus loading in a shallow eutrophic lake. Water Res 16: 911-919

Jassby A D Goldman C R (1974) Loss rates from a lake phytoplankton community. Limnol Oceanogr 19: 618-627

Jensen A (1984) Excretion of organic carbon as function of nutrient stress. In: Holm-Hansen O Bolis L Gilles R (eds) Marine phytoplankton and productivity. Springer, Berlin; Heidelberg, New York, pp 61-72

Jeppesen E Søndergaard M Mortensen E Kristensen P Riemann B Jensen H J Müller J P Sortkjær O Jensen J P Christoffersen K Bosselmann S Dall E (1990) Fish manipulation as a lake restoration tool in shallow, eutrophic temperate lakes 1: cross-analysis of three Danish case-studies. Hydrobiologia 200/201: 205-218

Kalff J Knoechel R (1978) Phytoplankton and their dynamics in oligotrophic and eutrophic lakes. Annu Rev Ecol Syst 9: 475-495

Kappers F I (1984) On population dynamics of the cyanobacterium *Microcystis aeruginosa*. PhD Thesis. University of Amsterdam, Amsterdam

Kerfoot W C Sih A (eds) (1987) Predation, direct and indirect impacts on aquatic communities. University Press of New England, Hanover

Kersting K (1983) Direct determination of the "threshold food concentration" for *Daphnia magna*. Arch Hydrobiol 96: 510-514

Kieber D J McDaniel J Mopper K (1989) Photochemical source of biological substrates in sea water: implications for carbon cycling. Nature 341: 637-639

Kilham P Hecky R E (1988) Comparative ecology of marine and freshwater phytoplankton. Limnol Oceanogr 33: 776-795

Kilham P Kilham S S (1980) The evolutionary ecology of phytoplankton. In: Morris I (ed) The physiological ecology of phytoplankton. Blackwell, Oxford, pp 571-592

Kilham S S (1987) Phytoplankton responses to changes in mortality rates. Verh Int Ver Limnol 23: 677-682

Kilham S S Kott C L Tilman D (1977) Phosphate and silicate kinetics for the Lake Michigan diatom *Diatoma elongatum*. J Great Lakes Res 3: 93-99

Knisely K Geller W (1986) Selective feeding of four zooplankton species on natural lake phytoplankton. Oecologia 69: 86-94

Kohl J-G Nicklisch A (1988) Ökophysiologie der Algen. Gustav Fischer, Stuttgart, 253 pp

Kooijman S A L M (1986) Population dynamics on the basis of budgets. In: Metz J A J Diekman O (eds) The dynamics of physiologically structured populations. Springer, Berlin, Heidelberg, New York, pp 266-297

Kring L R O'Brien W J (1976) Effect of varying oxygen concentration on the filtring rate of *Daphnia pulex*. Ecology 57: 808-814

Lampert W (1976) A directly coupled artificial two-step food chain for long term experiments with filter feeders at constant food concentrations. Mar Biol 37: 349-355

Lampert W (1977) Studies on the carbon balance of *Daphnia pulex* de Geer as related to environmental conditions. I - IV. Arch Hydrobiol Beih 48: 287-368

Lampert W (1978) Release of dissolved organic carbon by grazing zooplankton. Limnol Oceanogr 23: 831-834

Lampert W (1987) Feeding and nutrition in *Daphnia*. Mem Ist Ital Idrobiol 45: 143-192

Lampert W Bohrer R (1984) Effect of food availability on the respiratory quotient of *Daphnia magna*. Comp Biochem Physiol 78a: 221-224

Lampert W Gabriel W (1984) Tracer kinetics in *Daphnia*: an improved two-compartment model and experimental test. Arch Hydrobiol 100: 1-20

Lampert W Muck P (1985) Multiple aspects of food limitation in zooplankton communities: the *Daphnia–Eudiaptomus* example. Arch Hydrobiol Beih 21: 311-322

Lampert W Schober U (1980) The importance of "threshold" food concentrations. Am Soc Limnol Oceanogr Spec Symp 3: 264-267

Lampert W Fleckner W Rai H Taylor B E (1986) Phytoplankton control by grazing zooplankton: A study on the spring clear water phase. Limnol Oceanogr 31: 478-490

Langeland A Reinertsen H (1982) Interactions between phytoplankton and zooplankton in a fertilized lake. Holarct Ecol 5: 253-272

Lawless J F (1982) Statistical models and methods for lifetime data. Wiley, New York

Laws E A Bannister T T (1980) Nutrient- and light-limited growth of *Thalassiosira fluviatilis* in continuous culture, with implications for phytoplankton growth in the oceans. Limnol Oceanogr 25: 457-473

Laws E A Redalje D G Karl D M Chalup M S (1983) A theoretical and experimental examination of the predictions of two recent models of phytoplankton growth. J Theor Biol 105: 469-491

Laws E A Jones D R Terry K L Hirata J A (1985) Modifications in recent models of phyto-plankton growth: theoretical developments and experimental examination of predictions. J Theor Biol 114: 323-341

Lean D R S (1976) Movements of phosphorus between its biologically important forms in lake water. J Fish Res Board Can 33: 1525-1536

Lean D R S Nalewajko C (1976) Phosphate exchange and organic phosphorus excretion by freshwater algae. J Fish Res Board Can 33: 1312-1323

Le Borgne R (1982) Zooplankton production in the eastern tropical Atlantic Ocean: net growth efficiency and P:B in terms of carbon, nitrogen and phosphorus. Limnol Oceanogr 27: 681-698

Lee G F Rast W Jones R A (1978) Eutrophication of water bodies: insights for an age-old problem. Environ Sci Technol 12: 900-908

Lehman J T (1976) The filter-feeder as an optimal forager, and the predicted shapes of feeding curves. Limnol Oceanogr 21: 501-516

Lehman J T (1984) Grazing, nutrient release, and their impacts on the structure of phyto-plankton communities. In: Meyers D W Strickler J R (eds) Trophic interactions within aquatic ecosystems. Westview Press, Boulder, pp 49-72

Lehman J T (1988) Ecological principles affecting community structure and secondary production by zooplankton in marine and freshwater environments. Limnol Oceanogr 33: 931-945

Lehman J T Naumoski T (1985) Content and turnover of phosphorus in *Daphnia pulex*: Effect of food quality. Hydrobiologia 128: 119-125

Leon J A Thompson D B (1975) Competition between two species for two complementary or substitutable resources. J Theor Biol. 50: 185-201

Levin S A (1970) Community equilibria and stability, and an extension of the competitive exclusion principle. Am Nat 104: 413-423

Levins R (1979) Coexistence in a variable environment. Am Nat 114: 765-783

Lindeman R L (1942) The trophic-dynamic aspect of ecology. Ecology 23: 399-418

Lotka A J (1907) Relation between birth rates and death rates. Science 26: 21-22

Lotka A J (1925) Elements of physical biology. Dover Publications, New York

Løvstad Ø Wold T (1984) Determination of external concentrations of available phosphorus for phytoplankton populations. Verh Int Ver Limnol 22: 205-210

Luenberger D G (1979) Introduction to dynamic systems. Theory, models and applications. Wiley, New York, 446 pp

Lynch M (1977) Fitness and optimal body size in zooplankton populations. Ecology 58: 763-774

Lynch M (1980a) The evolution of cladoceran life histories. Q Rev Biol 55: 23-41

Lynch M (1980b) *Aphanizomenon* blooms: alternate control and cultivation by *Daphnia pulex*. Am Soc Limnol Oceanogr Spec Symp 3: 299-304

Lynch M (1983) Estimation of size-specific mortality rates in zooplankton populations by periodic sampling. Limnol Oceanogr 28: 533-545

Lynch M (1989) The life history consequences of resource depression in *Daphnia pulex*. Ecology 70: 246-256

Lynch M Ennis R (1983) Resource availability, maternal effects, and longevity. Exp Gerontol 18: 147-165

Lynch M Weider L J Lampert W (1986) Measurement of the carbon balance in *Daphnia*. Limnol Oceanogr 31: 17-33

MacArthur R H Wilson E O (1967) The theory of island biogeography. Princeton University Press, Princeton

MacDonald N (1978) Time lags in biological models. Springer, Berlin, Heidelberg, New York

Mackereth F J (1953) Phosphorus utilization by *Asterionella formosa*. J Exp Bot 4: 296-313

Malone T C (1980) Algal size.In: Morris I (ed) The physiological ecology of phytoplankton. Blackwell Scientific Publications, Oxford, pp 433-463

Mangel M Clark C W (1988) Dynamic modeling in behavioral ecology. Princeton University Press, Princeton, 308 pp

May R M (1972) Limit cycles in predator-prey communities. Science 177: 900-902

May R M (1975) Stability and complexity in model ecosystems. Princeton University Press, Princeton

May R M (1981) Models for two interacting populations. In: May R M (ed) Theoretical ecology. Principles and applications. Saunders, Philadelphia, pp 78-104

Mazumder A McQueen D J Taylor W D Lean D R S (1988) Effects of fertilization and planktovorous fish (yellow perch) predation on size-distribution of particulate phosphorus and assimilated phosphate: large enclosure experiments. Limnol Oceanogr 33: 421-430

Mazumder A Taylor W D McQueen D J Lean D R S (1989) Effects of fertilization and planktirous fish on epilimnetic phosphorus and phosphorus sedimentation in large enclosures. Can J Fish Aquat Sci 46: 1735-1742

Mazumder A McQueen D J Taylor W D Lean D R S (1990) Pelagic food web interactions and hypolimnetic oxygen depletion: Results from experimental enclosures and lakes. Aquat Sci 52: 144-155

Mazumder A Taylor W D Lean D R S McQueen D J (1992) Partitioning and fluxes of phosphorus: mechanisms regulating the size-distribution and biomass of plankton. Arch Hydrobiol Beih (Ergebn Limnol) 35: 121-143

McAllister C D LeBrasseur R J Parsons T R (1972) Stability of enriched aquatic ecosystems. Science 175: 562-564

McCauley E Kalff J (1981) Empirical relationships between phytoplankton and zooplankton biomass in lakes. Can J Fish Aquat Sci 38: 458-463

McCauley E Murdoch W W (1987) Cyclic and stable populations: plankton as a paradigm. Am Nat 129: 97-121

McCauley E Murdoch W W (1989) Predator-prey dynamics in environments rich and poor in nutrients. Nature 343: 455-457

McCauley E Murdoch W W Nisbet R M Gurney W S (1990) The physiological ecology of *Daphnia*: development of a model of growth and reproduction. Ecology 71: 703-715

McMahon J W (1965) Some physical factors influencing the feeding behaviour of *Daphnia magna* Straus. Can J Zool 43: 603-612

McMahon J W Rigler F H (1965) Feeding rate of *Daphnia magna* Straus in different foods labeled with radioactive phosphorus. Limnol Oceanogr 10: 105-113

McQueen D J Post J R Mills E L (1986) Trophic relationships in freshwater pelagic ecosystems. Can J Fish Aquat Sci 43: 1571-1581

McQueen D J Johannes M R S Post J R Stewart T J Lean D R S (1989) Bottom-up and top-down impacts on freshwater pelagic community structure. Ecol Monogr 59: 289-309

Menge B A Sutherland J P (1976) Species diversity gradients: synthesis of the roles of predation, competition, and temporal heterogeneity. Am Nat 110: 351-368

Menge B A Sutherland J P (1987) Community regulation: variation in disturbance, competition, and predation in relation to environmental stress and recruitment. Am Nat 130: 730

Metz J A J Diekman O (eds) (1986) The dynamics of physiologically structured populations. Springer, Berlin, Heidelberg, New York

Meusy J J (1980) Vitellogenin, the extraovarian precursor of the protein yolk in Crustacea: a review. Reprod Nutr Dev 83: 1-21.

Mickelson M J Maske H Dugdale R C (1979) Nutrient-dependent dominance in multispecies chemostat cultures of diatoms. Limnol Oceanogr 24: 298-315

Møller-Andersen J (1975) Influence of pH on release of phosphorus from lake sediments. Arch Hydrobiol 76: 411-419

Morel F M M (1987) Kinetics of nutrient uptake and growth in phytoplankton. J Phycol 23: 137-150

Morgan K C (1976) Studies on the autecology of the freshwater algal flagellate *Cryptomonas erosa*. PhD Thesis. McGill University, Montreal

Mortimer C H (1941) The exchange of dissolved substances between mud and water in lakes. I. J Ecol 29: 280-329

Mortimer C H (1942) The exchange of dissolved substances between mud and water in lakes. II. J Ecol 30: 147-201

Muck P Lampert W (1980) Feeding of freshwater filter-feeders at very low food concentrations: poor evidence for "threshold feeding" and "optimal foraging" in *Daphnia longispina* and *Eudiaptomus gracilis*. J Plankton Res 2: 367-379

Muck P Lampert W (1984) An experimental study on the importance of food conditions for the relative abundance of calanoid copepods and cladocerans. 1. Comparative feeding studies with Eudiaptomus gracilis and Daphnia longispina. Arch Hydrobiol Beih 66: 157-179

Müller V H (1972) Wachstum und Phosphatbedarf von *Nitzschia actinastroides* (Lem.) v. Goor in statischer und homocontinuierlicher Kultur unter Phosphatlimitierung. Arch Hydrobiol Beih 38: 399-484

Murdoch W W (1969) Switching in general predators: experiments on predator specificity and stability of prey populations. Ecol Monogr 39: 335-354

Murdoch W W McCauley E (1985) Three distinct types of dynamic behaviour shown by a single planktonic system. Nature 316: 628-630

Myers J Graham.J-R (1971) The photosynthetic unit in *Chlorella* measured by repetitive short flashes. Plant Physiol 48: 282-286

Myklestad S (1977) Production of carbohydrates by marine planktonic diatoms. II. Influence of the N/P ratio in growth medium on the assimilation ratio, growth rate, and production of cellular and extracellular carbohydrates by *Chaetoceros affinis* var. Willei (Gran) Hustedt and *Skeletonema costatum* (Grev.) Cleve. J Exp Mar Biol Ecol 29: 161-179

Nakashima B S Leggett W C (1980) Natural sources and requirements of phosphorus in fishes. Can J Fish Aquat Sci 37: 679-686

Nauwerck A (1963) Die Beziehungen zwischen Zooplankton und Phytoplankton im See Erken. Symb Bot Ups 17: 1-163

Nielsen M Olsen Y (1989) The dependence of the assimilation efficiency in *Daphnia magna* on the ^{14}C-labeling period of the food alga *Scenedesmus acutus*. Limnol Oceanogr 34: 1311-1315

Nisbet R M Gurney W S C (1982) Modelling fluctuating populations. Wiley, New York, 377 pp

Nisbet R M Gurney W S C Murdoch W W McCauley E (1989) Structured population models: a tool for linking effects at individual and population level. Biol J Linn Soc 37: 79-99

Nyholm N (1977) Kinetics of phosphate limited algal growth. Biotechnol Bioeng 19: 467-492

Odum E P (1962) Relationships between structure and function in the ecosystem. Jpn J Ecol 12: 108-118

Odum E P de la Cruz A A (1963) Detritus as a major component of ecosystems. Bull Am Inst Biol Sci 13: 39-40

Odum H T (1957) Trophic structure and productivity of Silver Springs, Florida. Ecol Monogr 27: 55-112

Oglesby R T (1977) Phytoplankton summer standing crops and annual productivity as functions of phosphorus loading and various physical factors. J Fish Res Board Can 34: 2255-2270

Oksanen L Fretwell S D Arruda J Niemela P (1981) Exploitation ecosystems in gradients of primary productivity. Am Nat 118: 240-261

Olsen Y (1988) Phosphate kinetics and competitive ability of planktonic blooming cyanobacteria under variable phosphate supply. Dr. Techn. Thesis. University of Trondheim, Trondheim

Olsen Y (1989) Evaluation of competitive ability of *Staurastrum luetkemuellerii* (Chlorophyceae) and *Microcystis aeruginosa* (Cyanophyceae) under P limitation. J Phycol 25: 486-499

Olsen Y Østgaard K (1985) Estimating release rates of phosphorus from zooplankton; model and experimental verification. Limnol Oceanogr 30: 844-852

Olsen Y Knutsen G Lien T (1983a) Characteristics of phosphorus limitation in *Chlamydomonas reinhardtii* (Chlorophyceae) and its palmelloids. J Phycol 19: 313-319

Olsen Y Jensen A Reinertsen H Rugstad B (1983b) Comparison of different algal carbon estimates by use of the Droop-model for nutrient limited growth. J Plankton Res 5: 43-51

Olsen Y Vårum K M Jensen A (1986a) Some characteristics of the carbon compounds released by *Daphnia*. J Plankton Res 8: 505-517

Olsen Y Jensen A Reinertsen H Børsheim K Y Heldal M Langeland A (1986b) Dependence of the rate of release of phosphorus by zooplankton upon the P:C ratio in the food supply, as calculated from the recycling-model. Limnol Oceanogr 31: 34-44

Olsen Y Vadstein O Andersen T Jensen A (1989) Competition between *Staurastrum luetkemuellerii* (Chlorophyceae) and *Microcystis aeruginosa* (Cyanophyceae) under varying modes of phosphate supply. J Phycol 25: 499-508

Paasche E (1980) Silicon. In: Morris I (ed) The physiological ecology of phytoplankton. Blackwell, Oxford, pp 258-284

Pace M L (1984) Zooplankton community structure, but not biomass, influences the phosphorus-chlorophyll a relationship. Can J Fish Aquat Sci 41: 1089-1096

Pace M L Porter K G Feig Y S (1983) Species- and age-specific differences in bacterial resource utilization by two co-occurrring cladocerans. Ecology 64: 1145-1156

Paloheimo J E Crabtree S J Taylor W D (1982) Growth model of *Daphnia*. Can J Fish Aquat Sci 39: 598-606

Parker R A Olson M I (1966) The uptake of inorganic phosphate by *Daphnia schoedleri*. Physiol Zool 39: 53-65

Pennak R W (1957) Species composition of limnetic zooplankton communities. Limnol Oceanogr 2: 222-232

Perrin N (1989) Reproductive allocation and size constraints in the cladoceran *Simocephalus vetulus* (Müller). Funct Ecol 3: 279-283

Perry M J (1976) Phosphate utilization by an oceanic diatom in phosphorus-limited chemostat culture and in the oligotrophic waters of the central North Pacific. Limnol Oceanogr 21: 88-107

Peters R H (1983) Ecological implications of body size. Cambridge University Press, Cambridge, New york, 329 pp

Peters R H (1987) Metabolism in *Daphnia*. Mem Ist Ital Idrobiol 45: 193-243

Peters R H Rigler F H (1973) Phosphorus release by *Daphnia*. Limnol Oceanogr 18: 821-839

Pianka E R (1970) On r- and K-selection. Am Nat 104: 592-597

Piyasiri S (1985) Methodological aspects of defining food dependence and food thresholds in fresh-water calanoids. Arch Hydrobiol Beih. (Ergebn Limnol) 21: 277-284

Pollard J H (1973) Mathematical models for the growth of human populations. Cambridge University Press, Cambridge, New york

Porter K G (1988) Phagotrophic protozoa in planktonic food webs. Hydrobiologia 159: 89-97

Porter K G Gerritsen J Orcutt J D (1982) The effect of food concentration on swimming patterns, feeding behaviour, ingestion, assimilation, and respiration by *Daphnia*. Limnol Oceanogr 27: 935-949

Porter K G Orcutt J D Gerritsen J (1983) Functional response and fitness in a generalist filter feeder, *Daphnia magna* (cladocera: crustacea). Ecology 64: 735-742

Powell T Richerson P J (1985) Temporal variation, spatial heterogeneity, and competition for resources in plankton systems: a theoretical model. Am Nat 125: 431-463

Prairie Y T (1988) A test of the sedimentation assumptions of phosphorus input-output models. Arch Hydrobiol 111: 321-327

Prairie Y T (1989) Statistical models for the estimation of net phosphorus sedimentation in lakes. Aquat Sci 51: 192-210

Press W H Flannery B P Teukolsky S A Vetterling W T (1986) Numerical recipes. The art of scientific computing. Cambridge University Press, Cambridge, 818 pp

Reckhow K H Chapra S C (1983) Engineering approaches for lake management, vol 1: Data analysis and empirical modeling. Butterworth, London

Redfield A C (1958) The biological control of chemical factors in the environment. Am Sci 46: 205-222

Reiners W A (1986) Complementary models for ecosystems. Am Nat 127: 59-73

Reinertsen H Jensen A Koksvik J I Langeland A Olsen Y (1989) Effects of fish removal on the limnetic ecosystem of a eutrophic lake. Can J Fish Aquat Sci 47: 166-173

Reynolds C S (1984) The ecology of freshwater phytoplankton. Cambridge University Press, Cambridge, 384 pp

Reynolds C S (1989) Physical determinants of phytoplankton succession. In: Sommer U (ed) Plankton Ecology. Springer, Berlin, Heidelberg, New York, pp 9-58

Rhee G-Y (1973) A continuous culture study of phosphate uptake, growth rate and polyphosphate in *Scenedesmus* sp. J phycol 9: 495-506

Rhee G-Y Gotham I J (1980) Optimum N:P ratios and coexistence of planktonic algae. J Phycol 16: 468-489

Rhee G-Y Gotham I J (1981) The effect of environmental factors on phytoplankton growth: light and the interaction of light with nitrate limitation. Limnol Oceanogr 26: 649-660

Richman S (1958) The transformation of energy by *Daphnia pulex*. Ecol Monogr 28: 273-291

Richman S Dodson S I (1983) The effect of food quality on feeding and respiration by *Daphnia* and *Diaptomus*. Limnol Oceanogr 28: 948-956

Ricker W E (1973) Linear regressions in fishery research. J Fish Res Board Can 30: 409-434

Riegman R Mur L R (1984) Regulation of phosphate uptake kinetics in *Oscillatoria agardhii*. Arch Microbiol 139: 28-32

Riemann B Søndergaard M (eds) (1986) Carbon dynamics in eutrophic, temperate lakes. Elsevier, Amsterdam

Rigler F H (1961) The relation between concentration of food and feeding rate of *Daphnia magna* Straus. Can J Zool 39: 857-868

Rigler F H (1964) The phosphorus fractions and turnover time of inorganic phosphorus in different types of lakes. Limnol Oceanogr 9: 511-514

Rodhe W (1978) Algae in culture and nature. Verh Int Ver Limnol 21: 7-20

Rogers T D (1981) Chaos in systems in population biology. Prog Theor Biol 6: 92-146

Rognerud S Kjellberg G (1984) Relationships between phytoplankton and zooplankton biomasses in large lakes. Verh Int Ver Limnol 22: 666-671

Rosenzweig M L (1969) Why the prey curve has a hump. Am Nat 103: 81-87

Rosenzweig M L (1971) Paradox of enrichment: Destabilization of exploitation ecosystems in ecological time. Science 171: 385-387

Rosenzweig M L MacArthur R H (1963) Graphical representation and stability conditions of predator-prey interactions. Am Nat 97: 209-223

Roughgarden J (1979) Theory of population genetics and evolutionary ecology: an introduction. Macmillan, New York. 634 pp

Sakamoto M (1966) Primary production by phytoplankton community in some Japanese lakes and its dependence on lake depth. Arch Hydrobiol 62: 1-28

Sakshaug E Andresen K Myklestad S Olsen Y (1983) Nutrient status of phytoplankton communities in Norwegian waters (marine, brackish, and fresh) as revealed by their chemical composition. J Plankton Res 5: 175-196

Sanni S Wærvågen S B (1990) Oligotrophication as a result of planktivorous fish removal with rotenone in the small, eutrophic lake, Mosvatn, Norway. Hydrobiologia 200/201: 263-274

Scavia D McFarland M J (1982) Phosphorus release patterns and the effects of reproductive stage and ecdysis in *Daphnia magna*. Can J Fish Aquat Sci 39: 1310-1314

Schaffer W M Kot M (1986) Chaos in ecological systems: the coals that Newcastle forgot. TREE 1: 58-63

Schindler D W (1977) Evolution of phosphorus limitation in lakes. Science 195: 260-262

Schindler D W (1978) Factors regulating phytoplankton production and standing crop in the world's freshwaters. Limnol Oceanogr 23: 478-486

Shapiro J (1980) The importance of trophic level interactions to the abundance and species composition of algae in lakes. Dev Hydrobiol 2: 105-116

Shapiro J Wright D I (1984) Lake restoration by biomanipulation:Round Lake, Minnesota, the first two years. Freshwater Biol 14: 371-383

Shapiro J Lamarra V Lynch M (1975) Biomanipulation: an ecosystem approach to lake restoration. In: Brezonik P L Fox J L (eds) Proceedings of a symposium on water quality management through biological control. University of Florida Press, Gainesville, pp 85-96

Sharp J H (1977) Excretion of organic matter by marine phytoplankton. Do healthy cells do it? Limnol Oceanogr 22: 381-399

Shuter B J (1978) Size dependence of phosphorus and nitrogen subsistence quotas in unicellular microorganisms. Limnol Oceanogr 23: 1248-1255

Shuter B (1979) A model of physiological adaption in unicellular algae. J Theor Biol 78: 519-552

Sinko J W Streifer W (1967) A new model for age-size structure of a population. Ecology 48: 910-918

Slagstad D (1981) Modeling and simulation of physiology and population dynamics of copepods. Effects of physical and biological parameters. Model Identif Control 2: 119-162

Slagstad D (1982) A model of phytoplankton growth – effects of vertical mixing and adaptation to light. Model Identif Control 3: 111-130

Smith F E (1963) Population dynamics in *Daphnia magna* and a new model for population growth. Ecology 44: 651-663

Smith R E H Kalff J (1982) Size-dependent phosphorus uptake kinetics and cell quota in phytoplankton. J Phycol 18: 275-284

Smith R E Kalff J (1983) Competition for phosphorus among co-occurring freshwater phytoplankton. Limnol Oceanogr 28: 448-464

Smith R E Kalff J (1985) Phosphorus competition among phytoplankton: A reply. Limnol Oceanogr 30: 440-444

Smith V H (1983) Low nitrogen to phosphorus ratios favor dominance by blue-green algae in lake phytoplankton. Science 221: 669-671

Sommer U (1981) The role of r- and K-selection in the succession of phytoplankton in Lake Constance. Acta Oecol 2: 327-342

Sommer U (1983) Nutrient competition between phytoplankton species in multispecies chemostat experiments. Arch Hydrobiol 96: 399-416

Sommer U (1984) The paradox of the plankton: fluctuations of the phosphorus availability maintain diversity of phytoplankton in flow-through cultures. Limnol Oceanogr 29: 633-636

Sommer U (1985) Comparison between steady state and nonsteady state competition: experiments with natural phytoplankton. Limnol Oceanogr 30: 335-346

Sommer U (1986) Phytoplankton competition along a gradient of dilution rates. Oecologia 68: 503-506

Sommer U (1988a) Does nutrient competition among phytoplankton occur in situ? Verh Int Ver Limnol 23: 707-712

Sommer U (1988b) Phytoplankton succession in microcosm experiments under simultaneous grazing pressure and resource limitation. Limnol Oceanogr 33: 1037-1054

Sommer U (1989a) Nutrient status and nutrient competition of phytoplankton in a shallow, hypertrophic lake. Limnol Oceanogr 34: 1162-1173

Sommer U (ed) (1989b) Plankton ecology. Springer, Berlin, Heidelberg, New York

Sommer U Kilham S S (1985) Phytoplankton natural community competition experiments: a reinterpretation. Limnol Oceanogr 30: 436-440

Sommer U Gliwicz Z M Lampert W Duncan A (1986) The PEG-model of seasonal succession of planktonic events in fresh waters. Arch Hydrobiol 106: 433-471

Souza-Machado S Rollins R Jacobs D Hartman J (1990) Studying chaotic systems using microcomputers and Lyapunov exponents. Am J Phys 58: 321-329

Steemann Nielsen E (1979) Growth of the unicellular alga *Selenastrum capricornutum* as a function of P. With some information also on N. Verh Int Ver Limnol 20: 36-42

Stemberger R S Gilbert J J (1985) Body size, food concentration, and population growth in planktonic rotifers. Ecology 66: 1151-1159

Stemberger R S Gilbert J J (1987) Rotifer threshold food concentrations and the size-efficiency hypothesis. Ecology 68: 181-187

Stenson J A E Bohlin T Henrikson L Nilsson B I Nyman H G Oscarson H G Larsson P (1978) Effects of fish removal from a small lake. Verh Int Ver Limnol 20: 794-801

Sterner R W (1989) The role of grazers in phytoplankton succession. In: Sommer U (ed) Plankton ecology. Springer, Berlin, Heidelberg, New York, pp 107-170

Sterner R W (1990) The ratio of nitrogen to phosphorus resupplied by herbivores: zooplankton and the algal competitive arena. Am Nat 136: 209-229

Sterner R W Hagemeier D D Smith W L Smith R F (1993) Phytoplankton nutrient limitation and food quality for *Daphnia*. Limnol Oceanogr 38: 857-871

Stross R G (1987) Photoperiodism and phased growth in *Daphnia* populations: coactions in perspective. Mem Ist Ital Idrobiol 45: 413-437

Suttle C A (1987) Effects of nutrient patchiness and N:P supply ratios on the ecology and physiology of freshwater phytoplankton. PhD Thesis, University British Columbia, Vancouver

Taylor B E (1985) Effects of food limitation on growth and reproduction of *Daphnia*. Arch Hydrobiol Beih 21: 285-296

Taylor W D Lean D R S (1991) Phosphorus pool sizes and fluxes in the epilimnion of a meso-trophic lake. Can J Fish Aquat Sci 48: 1293-1301

Tessier A J Goulden C E (1982) Estimating food limitation in cladoceran populations. Limnol Oceanogr 27: 707-717

Tessier A J Goulden C E (1987) Cladoceran juvenile growth. Limnol Oceanogr 32: 680-686

Tessier A J Henry L L Goulden C E Durand M W (1983) Starvation in *Daphnia*: energy reserves and reproductive allocation. Limnol Oceanogr 28: 667-676

Thingstad T F (1987) Analyzing the "microbial loop". PhD Thesis. University of Bergen, Bergen

Thompson J M Ferguson A J D Reynolds C S (1982) Natural filtration rates of zooplankton in a closed system: the derivation of a community grazing index. J Plankton Res 4: 545-560

Thompson J M T Stewart H B (1986) Nonlinear dynamics and chaos. Wiley, New York, 376 pp

Threlkeld S T (1987) *Daphnia* population fluctuation: patterns and mechanism. Mem Ist Ital Idrobiol 45: 367-388

Thurman E M (1985) Organic geochemistry of natural waters. Dr W Junk, The Hague

Tillmann U Lampert W (1984) Competitive ability of differently sized *Daphnia* species: an experimental test. J Freshwater Ecol 2: 311-323

Tilman D (1980) Resources: a graphical-mechanistic approach to competition and predation. Am Nat 116: 362-393

Tilman D (1981) Tests of resource competition theory using four species of Lake Michigan algae. Ecology 62: 802-815

Tilman D (1982) Resource competition and community structure. Princeton University Press, Princeton

Tilman D Kilham S S (1976b) Phosphate and silicate growth and uptake of the diatoms *Asterionella formosa* and *Cyclotella meneghiniana* in batch and semicontinuos culture. J Phycol 12: 375-383

Titman D Kilham P (1976b) Sinking in fresh water phytoplankton: some ecological implications of cell nutrient status and physical mixing processes. Limnol Oceanogr 21: 409-417

Tilman D Kilham S S Kilham P (1982) Phytoplankton community ecology: the role of limiting nutrients. Annu Rev Ecol Syst 13: 349-372

Tufillaro N B Abbott T Reilly J (1992) An experimental approach to nonlinear dynamics and chaos. Addison-Wesley, Redwood City, 340 pp.

Turner J T Ferrante J G (1979) Zooplankton fecal pellets in aquatic ecosystems. Bioscience 29: 670-677

Turpin D H (1988) Physiological mechanisms in phytoplankton resource competition. In: Sandgren C D (ed) Growth and reproductive strategies of freshwater phytoplankton. Cambridge University Press, Cambridge, pp 316-368

Uehlinger U (1980) Experimentelle Untersuchungen zur Autökologie von *Aphanizomenon flos-aquae*. Arch Hydrobiol Beih 60: 260-288

Vadstein O Olsen Y (1989) Chemical composition and phosphate uptake kinetics of limnetic bacterial communities cultured in chemostat under phosphorus limitation. Limnol Oceanogr 34: 939-946

Vadstein O Jensen A Olsen Y Reinertsen H (1988) Growth and phosphorus status of limnetic phytoplankton and bacteria. Limnol Oceanogr 33: 489-503

Van Donk E Gulati R D Grimm M P (1989) Food-web manipulation in Lake Zwemlust: positive and negative effects during the first two years. Hydrobiol Bull 23: 19-35

Vance R R (1978) Predation and resource partitioning in one predator-two prey model communities. Am Nat 112: 797-813

Vanderploeg H A Scavia D (1979) Calculation and use of selectivity coefficients of feeding: zooplankton grazing. Ecol Model 7: 135-149

Vanderploeg H A Scavia D Liebig J R (1984) Feeding rate of *Diaptomus sicilis* and its relation to selectivity and effective food concentration in algal mixtures and in Lake Michigan. J Plankton Res 6: 919-941

Van Liere L (1979) On *Oscillatoria agardhii* Gomont: experimental ecology and physiology of a nuisance bloom-forming cyanobacterium. PhD Thesis. University of Amsterdam, Amsterdam

Vanni M J Temte J (1990) Seasonal patterns of grazing and nutrient limitation of phytoplankton in a eutrophic lake. Limnol Oceanogr 35: 697-709

Vidal J (1980) Physioecology of zooplankton. 1. Effects of phytoplankton concentration, temperature, and body size on the growth rate of *Calanus pacificus* and *Pseudocalanus* sp. Mar Biol 56: 111-134

Vollenweider R A (1968) Scientific fundamentals of the eutrophication of lakes and flowing waters, with particular reference to nitrogen and phosphorus as factors in eutrophication. OECD, Paris

Vollenweider R A (1976) Advances in defining critical loading levels for phosphorus in lake eutrophication. Mem Ist Ital Idrobiol 33: 53-83

Volterra V (1926) Variations and fluctuations of the number of individuals in animal species living together. J Cons Perm Int Ent Mer 3: 3-51

Walsby A E Reynolds C S (1980) Sinking and floating. In: Morris I (ed) The physiological ecology of phytoplankton. Blackwell Scientific Publications, Oxford, pp 371-412

Wetzel R G (1975) Limnology. Saunders, Philadelphia, 743 pp

Wetzel R G Rich Ph Miller C H Allen H L (1972) Metabolism of dissolved and particulate detrital carbon in a temperate hard-water lake. Mem Ist Ital Idrobiol 29: 185-243

Williamson C E Gilbert J J (1980) Variation among zooplankton predators: The potential of *Asplanchna*, *Mesocyclops*, and *Cyclops* to attack, capture, and eat various rotifer prey. Am Soc Limnol Oceanogr Spec Symp 3: 509-517

Williamson C E Butler N M Forcina L (1985) Food limitation in naupliar and adult *Diaptomus pallidus*. Limnol Oceanogr 30: 1283-1290

Winner R W Farrell M P (1976) Acute and chronic toxicity of copper to four species of *Daphnia*. J Fish Res Board Can 33: 1685-1691

Wright D I Shapiro J (1984) Nutrient reduction by biomanipulation: an unexpected phenomenon and its possible cause. Verh Int Ver Limnol 22: 518-524

Yan N D (1986) Empirical prediction of crustacean zooplankton biomass in nutrient-poor Canadian shield lakes. Can J Fish Aquat Sci 43: 788-796

Yan N D Mackie G L Boomer D (1989) Seasonal patterns in metal levels of the net plankton of three canadian shield lakes. Sci Total Environ 87/88: 439-461

Zaffagnini F (1987) Reproduction in *Daphnia*. Mem Ist Ital Idrobiol 45: 245-284

Zaret T M (1980) Predation and freshwater communities. Yale University Press, New Haven

Zeuthen E (1953) Oxygen uptake as related to body size in organisms. Q Rev Biol 28: 1-12

Zevenboom W Mur L R (1980) N_2-fixing cyanobacteria: why they do not become dominant in Dutch hypertrophic lakes? In: Barica J Mur L R (eds) Hypertrophic ecosystems. Dr W Junk, The Hague, pp 123-130

Appendices

The following sections contain material that should not be necessary to read in order to follow the general line of reasoning in the preceding chapters, but which may still be of importance to those that are specially interested in some particular topic.

A1 Coexistence in a Gradient of N:P Supply Ratios

If we consider the equilibrium situation for species i alone (that is, $\mu_i = D$), we have from Eqs. (3.18), (3.19) that uptake and growth must balance (i.e., $v_{R,i} = \mu_i Q_{R,i}$) for both N and P. If growth is limited by element R, this means that all state variables and process rates for this element will attain their equilibrium values for this particular loss rate, which can be written as $\mu_{R,i} = D$, $S_R = S^*_{R,i}$, $Q_{R,i} = Q^*_{R,i}$, $\alpha_{R,i} = \alpha^*_{R,i}$, etc. On the other hand, if growth is not limited by element R, we must have $\mu_{R,i} > D$, $S_R > S^*_{R,i}$, $Q_{R,i} > Q^*_{R,i}$ and $\alpha_{R,i} < \alpha^*_{R,i}$.

If we identify the abstract resources, labeled I and II in Fig. 3.8, as concentrations of dissolved inorganic P and N (S_P and S_N), the isoclines for species i in Fig. 3.8 will be formed by the intersection of the two lines $S_P = S^*_{P,i}$ and $S_N = S^*_{N,i}$. If we maintain the arrangement of isoclines for species 1 and 2 as in Fig. 3.8, a necessary condition for coexistence will be that the isoclines intersect; that is, species 1 is the superior competitor for P ($S^*_{P,1} < S^*_{P,2}$), and species 2 is the superior competitor for N ($S^*_{N,1} > S^*_{N,2}$). Since the maximal growth rates are assumed to be identical ($\mu''_1 = \mu''_2 = \mu''$), the condition $S^*_{R,1} < S^*_{R,2}$ becomes equivalent to $K'_{R,1} < K'_{R,2}$, where $K'_{R,i}$ [(μg R) l^{-1}] is the Monod half-saturation parameter for species i when limited by element R. By substituting the definition of $K'_{R,i}$ [Eq. (3.11)], the conditions for intersecting isoclines can be written as:

$$\frac{\alpha'_{P,2}}{\alpha'_{P,1}} < \frac{Q'_{P,2}}{Q'_{P,1}} \tag{A1.1}$$

$$\frac{\alpha'_{N,2}}{\alpha'_{N,1}} > \frac{Q'_{N,2}}{Q'_{N,1}}. \tag{A1.2}$$

When the isoclines intersect, it becomes possible for the two species to coexist in a situation where each species is limited by the nutrient on which it is the inferior competitor (i. e., species 1 is limited by N and species 2 by P). In order to have a stable equilibrium at the isocline intersection point $(S^*_{P,2}, S^*_{N,1})$ we must, also require that each species consumes relatively more of the nutrient that it is limited by. In terms of nutrient uptake rates this condition can be written as $v_{N,1}/v_{P,1} > v_{N,2}/v_{P,2}$, or equivalently, by substituting Eq. (3.20):

$$\frac{\alpha^*_{N,1} S^*_{N,1}}{\alpha^*_{P,1} S^*_{P,2}} > \frac{\alpha_{N,2} S^*_{N,1}}{\alpha^*_{P,2} S^*_{P,2}}, \tag{A1.3}$$

which becomes

$$\frac{\alpha^*_{N,1}}{\alpha^*_{P,1}} > \frac{\alpha_{N,2}}{\alpha^*_{P,2}} \tag{A1.4}$$

after eliminating $S^*_{P,2}$ and $S^*_{N,1}$ from Eq. (A1.3). Since species 1 is assumed not to be limited by P at the isocline intersection point, we will have $\alpha_{P,1} < \alpha^*_{P,1}$, and likewise, since species 2 is assumed not to be N-limited, $\alpha_{N,1} < \alpha^*_{N,1}$. In other words, Eq. (A1.4) will certainly be satisfied if we assume that

$$\frac{\alpha^*_{N,1}}{\alpha^*_{P,1}} > \frac{\alpha^*_{N,2}}{\alpha^*_{P,2}}. \tag{A1.5}$$

From Eq. (3.7) we know that we can express the equilibrium uptake affinity as a function of the steady state growth rate ($\mu_i = D$) by:

$$\alpha^*_{R,i} = \alpha'_{R,i} \frac{\mu'_{R,i}}{\mu''} \frac{\mu'' - D}{\mu'_{R,i} - D}. \tag{A1.6}$$

Substituting Eq. (A1.6) into Eq. (A1.5) yields

$$\left[\frac{\mu'_{N,1}}{\mu'_{P,1}} \frac{\mu'_{P,1} - D}{\mu'_{N,1} - D}\right]\frac{\alpha'_{N,1}}{\alpha'_{P,1}} > \left[\frac{\mu'_{N,2}}{\mu'_{P,2}} \frac{\mu'_{P,2} - D}{\mu'_{N,2} - D}\right]\frac{\alpha'_{N,2}}{\alpha'_{P,2}}. \tag{A1.7}$$

If the two species have equal storage capacities (that is, $Q''_{R,1}/Q'_{R,1} = Q''_{R,2}/Q'_{R,2}$) for each nutrient, then Eq. (3.23) implies that $\mu'_{R,1} = \mu''_{R,2}$ (notice that this does not exclude the possibility of having $\mu'_{P,1} \neq \mu'_{N,2}$). Under the assumption of equal storage capacities, the inequality Eq. (A1.7) can therefore be further simplified to:

$$\frac{\alpha'_{P,2}}{\alpha'_{P,1}} > \frac{\alpha'_{N,2}}{\alpha'_{N,1}}. \tag{A1.8}$$

A2 Allocation Rates and Growth History Data

In a typical *Daphnia* life table experiment, a cohort of animals are reared in separate vessels, and individual growth and reproduction are monitored from birth to death. While reproduction can be quantified directly by collecting produced offspring, body growth can only be measured nondestructively by indirect methods, such as carapace length measurements and length-weight regressions. As eggs are extruded and carapace is expanded simultaneously at molt time in cladocerans, neither reproduction nor somatic growth can be observed between instar transitions.

The growth history of a single cohort member in a life table experiment can be represented by a sequence of instar durations $(D_1, D_2,...)$ and the corresponding sequences of carbon budget terms for each instar:

$B_1, B_2,...$ body mass at the start of each instar.
$E_1, E_2,...$ egg mass produced during each instar.
$M_1, M_2,...$ body mass lost by molting at the end of each instar.

While the molting losses in principle could be observed directly be collecting cast-off exuviae, practice has been to estimate molt size indirectly from carapace length [Eqs. (4.6), (4.7)].

The individual terms of the carbon budget for each instar together define the net production rate $(F_i; (\mu g\ C)\ day^{-1})$ within instar i

$$F_i = D_i^{-1} \left(\left((B_{i+1} + M_i) - B_i \right) + E_i \right), \tag{A2.1}$$

and the fraction of net assimilate invested into reproduction during the same instar:

$$R_i = E_i \left(\left((B_{i+1} + M_i) - B_i \right) + E_i \right)^{-1}. \tag{A2.2}$$

The continuous equivalent to Eq. (A2.1), with F as a function of the time spent in instar i can be written as

$$F = D_i^{-1} \left((B - B_i) + E \right), \tag{A2.3}$$

with B and E given by the solution of Eqs. (4.3) and (4.4), such that $F(0) = 0$ and $F(D_i) = F_i$. Differentiating Eq. (A2.3) with respect to time and inserting Eqs. (4.3) and (4.4), gives

$$\dot{F} = D_i^{-1} \left(\dot{B} + \dot{E} \right) = D_i^{-1} g\ B. \tag{A2.4}$$

If we assume that the net assimilation rate can be approximated by a constant $g = g_i$ in instar i, the solution of Eq. (A2.4) can be written as

$$F_i = g_i\ D_i^{-1} \int_0^{D_i} B(\tau) d\tau. \tag{A2.5}$$

If body growth is exponential within instar i (but with a rate constant different from the net assimilation rate when the animal is reproducing), the body mass at any given time $t \in [0, D_i]$ still be given by

$$B(\tau) = B_i \exp\left(\left(D_i^{-1} \ln\left(\frac{B_{i+1} + M_i}{B_i}\right)\right)\tau\right) \tag{A2.6}$$

Inserting Eq. (A2.6) into Eq. (A2.5), integrating, and solving with respect to g_i, gives the relationship between the specific net assimilation rate of the continuous model [Eqs. (4.3) and (4.4)], and the carbon budget terms from the observed growth history:

$$g_i = D_i^{-1} \frac{((B_{i+1} + M_i) - B_i) + E_i}{(B_{i+1} + M_i) - B_i} \ln\left(\frac{B_{i+1} + M_i}{B_i}\right). \tag{A2.7}$$

In nonreproducing animals, where $E_i = 0$, the net assimilation rate estimate [Eq. (A2.7)] will be identical to $g_i = D_i^{-1} \ln((B_{i+1} + M_i)/B_i)$, the exponential growth rate estimate. On the other hand, when the investment in body growth is small ($B_{i+1} \approx B_i$), the first-order approximation $\ln x \approx x - 1$ can be used to show that $g_i \approx F_i / B_i$.

A3 Elemental Composition and Allocation Constraints

If we consider a single organism containing quantities c_1 and c_2 (µg ind.$^{-1}$) of two essential elements, the ratio of the two elements will be given by $\theta = c_1/c_2$. For this organism to maintain constant elemental ratios we have to make the requirement that $\dot{\theta} = 0$, which by application of the rule for the derivative of a quotient is equivalent to

$$\dot{c}_1 c_2 = c_1 \dot{c}_2. \tag{A3.1}$$

This is just a formal way of stating that maintaining constant elemental ratios requires the net rates of loss or gain in two elements to be balanced to the body proportions of the same two elements. The net rate of change of element j ($j = 1, 2$) can, in analogy with Eq. (4.2), be written as

$$\dot{c}_j = (\varepsilon_j I_j - r_j)c_j. \tag{A3.2}$$

The gains of element j are determined by the assimilation efficiency ε_j and the specific ingestion rate I_j, while the combined losses through excretion, secretion, or respiration (in the case of carbon) are contained in the specific loss rate r_j (all rates in units of day^{-1}). Inserting Eq. (A3.2) into Eq. (A3.1) cancels out the product $c_1 c_2$ on both sides of the equation, such that the requirements of balanced growth can be stated as

$$\varepsilon_1 I_1 - r_1 = \varepsilon_2 I_2 - r_2. \tag{A3.3}$$

If food particles are caught as whole entities, as in filter-feeding zooplankton, we can assume that ingestion of element j is related to the ambient concentration C_j [(μg) l^{-1}] of this element in the form of ingestible food particles as

$$I_j c_j = f C_j, \tag{A3.4}$$

where f is clearance rate of the organism (l ind.$^{-1}$ day^{-1}). From the two equations corresponding to $j = 1, 2$ in Eq. (A3.4), we can eliminate the clearance rate f, which must be the same for both elements. The element-specific ingestion rates of the two elements must then be related through

$$I_1 = (Q/\theta) I_2, \tag{A3.5}$$

where $Q = C_1/C_2$ is the elemental ratio of the food particles. Inserting Eq. (A3.5) into Eq. (A3.3) gives the requirement for balanced growth as

$$\varepsilon_1 (Q/\theta) I_2 - r_1 = \varepsilon_2 I_2 - r_2. \tag{A3.6}$$

So far, the model is completely general in that it describes balanced growth on any two elements, but not very useful in that five parameters in Eq. (A3.6) are left unspecified. To be more specific, we will for the rest of this section assume that element 2 is carbon, the major constituent and energy source in living organisms, while element 1 is phosphorus, the mineral nutrient most likely to become limiting to freshwater zooplankton. The parameters θ and Q will then be the P:C ratio [(μg P) (mg C)$^{-1}$] of the animal and its food, respectively. If we substitute the indices 1 and 2 with P and C and drop the index on the ingestion rate I, Eq. (A3.6) can be written as

$$\varepsilon_P (Q/\theta) I - r_P = \varepsilon_C I - r_C, \tag{A3.7}$$

where ε_P and ε_C are the assimilation efficiencies of ingested P and C while r_P and r_C equal the specific rates of P excretion and C respiration (day^{-1}). The P excretion rate is the flux resulting from the catabolism of organic phosphate esters, while the total flux of recycled P from the animal is the sum of excretion and egestion of P in unassimilated food. The specific ingestion rate I (day^{-1}) will be determined by the mechanics of filter-feeding, and can therefore be treated as an external parameter in the same way as Q and θ. If r_C also is considered a fixed rate determined by the energetic requirements of the animal, we end up with three free parameters ($\varepsilon_P, \varepsilon_C$ and r_P) which the animal must regulate in such a way that they satisfy Eq. (A3.7) in order to maintain a constant P:C ratio.

It is obvious that there must exist upper limits ε_P^* and ε_C^* such that the assimilation efficiencies of food P and C are constrained to the ranges $0 \leq \varepsilon_P \leq \varepsilon_P^* \leq 1$ and $0 \leq \varepsilon_C \leq \varepsilon_C^* \leq 1$. Furthermore, the phosphorus excretion rate must be non-negative: $r_P \geq 0$, as the contrary would imply direct uptake of dissolved P from the water phase. Within these constraints, there are still infinitely many ways in which animals may adjust the free parameters to conform with Eq. (A3.7).

It is also reasonable that there should be an optimal food composition Q^* that allows maximum utilization of both P and C in the food; that is, $\varepsilon_p = \varepsilon^*_p$ and $\varepsilon_c = \varepsilon^*_c$ when $Q = Q^*$. Among possible models describing adjustment to non-optimal food composition, a particularly simple formulation results from assuming a switching mechanism similar to the Liebig law of the minimum: *assimilation efficiency is maximal for the element least in supply.* This can be stated more formally as: $Q \geq Q^*$ implies that $\varepsilon_p \leq \varepsilon^*_p$ and $\varepsilon_c = \varepsilon^*_c$, while $Q \leq Q^*$ implies that $\varepsilon_p = \varepsilon^*_p$ and $\varepsilon_c \leq \varepsilon^*_c$.

If we focus on the effects on zooplankton growth of a suboptimal food P content, we need only to consider the case $Q \leq Q^*$. Under the assumption that P assimilation is maximal when P is least in supply, Eq. (A3.7) can be written as

$$\varepsilon_c I - r_c = \varepsilon^*_p(Q/\theta)I - r_p \leq \varepsilon^*_p(Q^*/\theta)I - r_p, \qquad (A3.8)$$

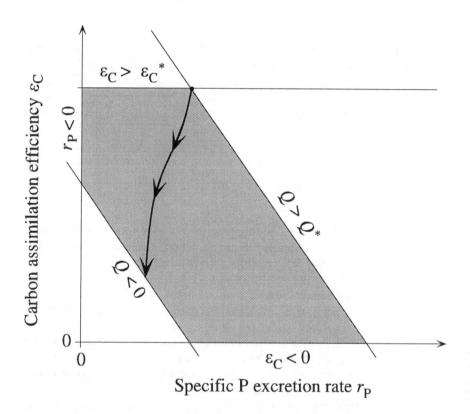

Fig. A3.1. Feasible region (*shaded area*) resulting from the constraints placed on phosphorus excretion rate r_p and carbon assimilation efficiency ε_c in order to maintain balanced growth on P-deficient food. *Directed path* shows an arbitrary example of a valid strategy for maintaining balanced growth when food P content decreases from optimal to zero

where the last inequality comes from the condition $Q \leq Q^*$. Likewise, we can use the non-negativity of the food P:C ratio ($Q \geq 0$) to obtain an additional constraint

$$\varepsilon_C I - r_C = \varepsilon_P^*(Q/\theta)I - r_P \geq -r_P. \tag{A3.9}$$

The two inequalities [Eqs. (A3.8), (A3.9)], and the necessary conditions $0 \leq \varepsilon_C \leq \varepsilon_C^* \leq 1$ and $r_P \geq 0$, define a feasible region containing all combinations of carbon assimilation efficiency ε_C and P excretion rate r_P that are compatible with Eq. (A3.7) when food is phosphorus deficient ($0 \leq Q \leq Q^*$). In Fig. A3.1 all trajectories confined within the feasible region, starting at the intersection of the lines corresponding to $\varepsilon_C = \varepsilon_C^*$ and $Q = Q^*$, and ending on the line where $Q = 0$, will describe a valid strategy for maintaining balanced growth when food P content decreases from Q^* to 0.

A particularly simple set of alternative food utilization strategies can be formulated if we can assume that there is a common upper limit to the assimilation efficiencies of both elements (for example determined by the maximum gut residence time), such that and if we assume that the optimal food composition is equal to the P:C ratio of the grazer; that is, $Q^* = \theta$. Under these conditions, the requirements of balanced growth [Eq. (A3.7)] on P-deficient food ($Q^* \leq \theta$) can be written as

$$\varepsilon_C I - r_C = \varepsilon^*(Q/\theta)I - r_P. \tag{A3.10}$$

At the optimal food composition, where $Q = \theta$ and $\varepsilon_P = \varepsilon_C = \varepsilon^*$, Eq. (A3.10) implies that $r_P = r_C$. Although such strategies can be contained within the feasible region of Fig. A3.1, it seems unnecessarily wasteful if P excretion increases such that $r_P > r_C$, when food becomes P deficient ($Q < \theta$). If we assume that $r_P \leq r_C$, then the family of power functions

$$r_{P,n} = r_C(Q/\theta)^n \tag{A3.11}$$

are able to describe an interesting range of possible P utilization strategies. Fig. A3.2 shows that as the exponent n decreases towards 0, the P excretion rate approaches the limiting case where $r_P = r_C$ for all $Q \leq \theta$. At the other extreme, when the exponent n increases towards infinity, the P excretion rate approaches the limiting case where $r_P = 0$ for all $Q < \theta$. The exponent n might therefore be interpreted as the animal's ability to economize with P, by reclaiming catabolites for anabolic purposes. The limiting cases $n \to 0$ and $n \to \infty$ would then correspond to no and complete reutilization of catabolites.

If we substitute Eq. (A3.11) into the growth-balancing equation [Eq. (A3.10)], the relationship between food P content and C assimilation efficiency can be written as

$$\varepsilon_{C,n} = \varepsilon^*\left((Q/\theta) + (1 - K_2^*)(1 - (Q/\theta)^n)\right), \tag{A3.12}$$

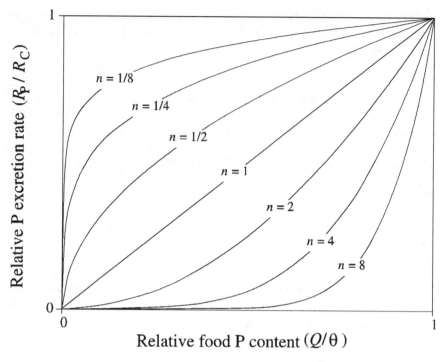

Fig. A3.2. A family of power functions, given by Eq. (A3.11), showing possible relationships between food P content and P excretion rate for different values of the exponent n

where we have also introduced a dimensionless ratio K_2^* that might be called the maximal net growth efficiency under optimal food composition:

$$K_2^* = \frac{\varepsilon^* I - r_C}{\varepsilon^* I}. \tag{A3.13}$$

Figure A3.3 shows that the relationship between C assimilation efficiency and food P content, for all values of the exponent n, is confined to a narrow band delimited by the extremes of wastefulness and conservation corresponding to $n = 0$ and $n = \infty$ in Eq. (A3.12). It thus seems reasonable that the models corresponding to $n = 0$, 1, and ∞ in Eq. (A3.12) constitute a representative subset of the strategies that might be found in animals maintaining a constant P : C ratio under suboptimal food composition. If we take into account that $(Q/\theta)^n \to 0$ as $n \to \infty$ if $Q < \theta$, and $(Q/\theta)^n \to 1$ as $n \to 0$ if $Q > \theta$, the three strategies can be written as

$$\varepsilon_{C,0} = \varepsilon^* (Q/\theta), \tag{A3.14}$$

$$\varepsilon_{C,1} = \varepsilon^* \left(1 - K_2^* (1 - Q/\theta)\right), \text{ and} \tag{A3.15}$$

$$\varepsilon_{C,\infty} = \varepsilon^* \left(1 - \left(K_2^* - Q/\theta\right)_+\right). \tag{A3.16}$$

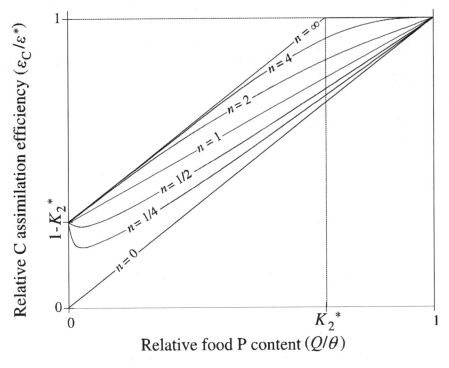

Fig. A3.3. Food C assimilation efficiency as function of food P content, as predicted by Eq. (A3.12), for different values of the exponent n

To ensure that $\varepsilon_{C,\infty} \le \varepsilon^*$, Eq. (A3.16) is formulated such that $\varepsilon_{C,\infty} = \varepsilon^*$ when $Q/\theta \ge K'_2$ [the operator $()_+$ is zero whenever the expression inside the parenthesis is ≤ 0].

A4 Survival Curves and Mortality Rates

Individual mortality can be modeled as a stochastic process with the conditional probability of dying within a given age increment Δx given by

$$\Pr\{\text{dying in } (x, x + \Delta x] | \text{ alive at } x\} = \eta(x)\Delta x. \qquad \text{(A4.1)}$$

The function $\eta(x)$ is the instantaneous mortality rate (day^{-1}), or the force of mortality (Lawless 1982) which is here assumed to be dependent on age alone. If a single cohort, initially consisting of n_0 members, is followed through time, the number of individuals alive at age $x + \Delta x$ will be equal to the number of survivors to age x, minus the cohort members that die in the interval $(x, x + \Delta x]$. This can be written as

$$n(x + \Delta x) = n(x) - \eta(x)\Delta x\, n(x). \tag{A4.2}$$

It is easy to show (see e.g., Nisbet and Gurney 1982) by rearranging Eq. (A4.2), that in the limit, as $\Delta x \to 0$, Eq. (A4.2) is equivalent to the differential equation

$$\dot{n}(x) = -\eta(x)n(x) \tag{A4.3}$$

with the initial condition $n(0) = n_0$. The solution to Eq. (A4.3) can be written as

$$n(x) = n_0\, e^{-\int_0^x \eta(\xi)\, d\xi}. \tag{A4.4}$$

If we define the fraction of the initial cohort surviving to age x as $l(x) = n(x)/n_0$, we obtain what is usually called the survival function of the cohort

$$l(x) = e^{-\int_0^x \eta(\xi)\, d\xi}. \tag{A4.5}$$

The survival function is related to the unconditional probability of an individual dying before reaching the age of x, $\Pr\{\text{dying before age } X \le x\} = F(x)$, by

$$F(x) = 1 - l(x). \tag{A4.6}$$

The probability density function of individual lifetimes will then be given by

$$f(x) = \dot{F}(x) = -\dot{l}(x), \tag{A4.7}$$

and the average individual lifetime $\bar{x}$, by

$$\bar{x} = \int_0^\infty x\, f(x)\, dx = -\int_0^\infty x\, \dot{l}(x)\, dx = \int_0^\infty l(x)\, dx, \tag{A4.8}$$

where the last identity comes from integration by parts. In other words, the integral of the survival function is the average life span, or the longevity, of individuals in the population. By comparison, the median individual lifetime $x_{\frac{1}{2}}$, is given by

$$x_{\frac{1}{2}} = F^{-1}\!\left(\tfrac{1}{2}\right) = l^{-1}\!\left(\tfrac{1}{2}\right). \tag{A4.9}$$

A cohort will die off exactly as described by Eq. (A4.4) only in the limiting case when $n_0 \to \infty$; the observed survival curve in a cohort with a finite initial size n_0 will be the result the random deaths of its cohort members, and, as such, a stochastic variable itself. It can be shown (e.g., Lawless 1982) that $N(x)$, the number of individuals being alive at age x from an initial cohort of n_0 members, will be a binomially distributed random variable. The probability of observing $N(x) = n$ at age x will therefore be given by

$$\Pr\{N(x) = n\} = \binom{n_0}{n} (l(x))^n (1 - l(x))^{n_0 - n}, \tag{A4.10}$$

with the expectation and variance of $N(x)$ being

$$E(N(x)) = n_0 l(x) \text{ and} \qquad\qquad\qquad\qquad (A4.11)$$

$$\text{Var}(N(x)) = n_0 l(x)(1 - l(x)). \qquad\qquad\qquad (A4.12)$$

It is seen from Eq. (A4.12) that the variance in the observed survival curve $N(x)$, will be maximal at the median survival time, that is, when $l(x) = \frac{1}{2}$. It is also seen from Eqs. (A4.11), (A4.12) that $N(x)/n_0$ is an unbiased estimator of true survival function $l(x)$, and that it is consistent (the variance of decreases to zero as $n_0 \to \infty$). A $1 - \alpha$ confidence interval for the number of survivors at age x will be given by numbers n' and n'' satisfying

$$\sum_{n=0}^{n=n'} \Pr\{N(x) = n\} = \alpha/2 \text{ and} \qquad\qquad\qquad (A4.13)$$

$$\sum_{n=0}^{n=n''} \Pr\{N(x) = n\} = 1 - \alpha/2 \cdot \qquad\qquad\qquad (A4.14)$$

Due to the amount of work involved in a typical life-table experiment, most *Daphnia* survival studies have followed only a single cohort per treatment, with a relatively small initial cohort size (20-40 individuals). A notable exception is the work of Frank et al. (1957), who used initial cohorts of up 800 *Daphnia pulex* individuals, with up to eight replicates within each treatment (see Section 4.4 for more details on their experimental design). Frank et al. (1957) reported the standard deviation of the observed survival curve for each treatment, which can be compared to the predicted standard deviation from the binomial model Eq. (A4.12):

$$S_{l(x)} = \sqrt{n_0^{-1} l(x)(1 - l(x))}. \qquad\qquad\qquad (A4.15)$$

As Frank et al. (1957) used a constant culture volume of 25 ml, the initial cohort size n_0 in their experiments will be directly proportional to animal density in each treatment (1-32 ml^{-1}).

Figure A4.1 shows that the standard deviation of the survival curve has a general resemblance to the concave relationship predicted by Eq. (A4.15), although the actual variability among replicated survival experiments is, for most treatments, much higher than would be expected from the binomial model. This means that the binomial confidence limits Eqs. (A4.13), (A4.14) should be regarded as very optimistic, and that other sources of variance can contribute strongly to the total variability among replicated survival experiments, even within the same laboratory.

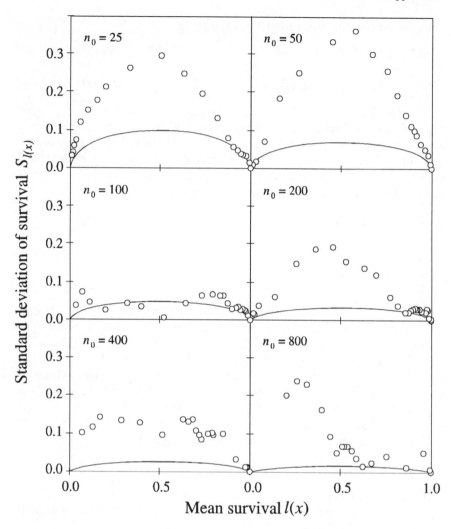

Fig. A4.1. Variability in *Daphnia* survival curves. *Open symbols* are observed standard deviations within each crowding level, as functions of mean survival. (Data from Frank et al. 1957). *Solid lines* are the predicted relationships between mean and standard deviation from Eq. (A4.15) for the same initial cohort sizes

A5 Elements of Age Distribution Theory

Assume that a population at a given time t (days) can be described by a continuous distribution $n(t,x)$, where x is individual age (days), such that the total number of individuals in the population $N(t)$ is given by

$$N(t) = \int_0^\infty n(t,x)dx \cdot \qquad (A5.1)$$

The individuals of age x at time t must obviously have been born at time $t - x$, but only a fraction $l(x)$ (see Appendix A4) of this cohort will be alive at time t:

$$n(t,x) = l(x)\, n(t - x, 0) \cdot \qquad (A5.2)$$

If we define the maternity function $m(x)$ (day^{-1}) as the number of offspring produced by a female of age x in a unit of time, then the number of newborn at time t will be the sum of births of all females:

$$n(t,0) = \int_0^\infty m(x)n(t,x)dx = \int_0^\infty m(x)l(x)n(t - x, 0)dx, \qquad (A5.3)$$

where the last identity comes from the substitution of Eq. (A5.2). This integral equation, which is often called the renewal condition of the population, relates the number of newborn at a given instant to the whole prehistory of births in the population (back to $t = -\infty$, to be exact!).

If the mortality and fecundity schedules [$l(x)$ and $m(x)$] remain constant over time, it can be shown (e.g., Pollard 1973 or Frauenthal 1986) that $n(t, x)$ converges to a stable age distribution with the number of newborn increasing exponentially as

$$n(t,0) = n_0\, e^{\lambda t} \cdot \qquad (A5.4)$$

It should be pointed out that the asymptotic proportionality factor n_0 will generally be different from the initial number of newborn in a population started from an arbitrary age distribution. By substituting Eq. (A5.4) into Eq. (A5.3), it is seen that the asymptotic population growth rate λ (day^{-1}), often called the intrinsic rate of increase, is given by the single real root (e.g., Pollard 1973) of the characteristic equation:

$$\int_0^\infty m(x)l(x)e^{-\lambda x}\, dx = 1 \cdot \qquad (A5.5)$$

The net reproductive value R_0, or the expected number of offspring to be produced in the lifetime of a newborn female (Frauenthal 1986), is given by

$$R_0 = \int_0^\infty m(x)l(x)dx \cdot \qquad (A5.6)$$

By comparing Eqs. (A5.5) and (A5.6), it is seen that in a population with zero asymptotic net growth rate ($\lambda = 0$), the net reproductive value of a female will be $R_0 = 1$. That is; each female is able to produce only the single offspring that is necessary to take her place when she dies, in order to maintain population size.

By substituting Eqs. (A5.2) and (A5.4) into (A5.1), it is seen that when the population has converged to the stable age distribution, total population size will be given by

$$N(t) = \int_0^\infty l(x)n(t-x,0)dx = n_0 e^{\lambda t} \int_0^\infty l(x)e^{-\lambda x} dx. \qquad (A5.7)$$

Already Lotka (1907) noticed that the specific birth rate β (day^{-1}) of the stable age distribution can be found from combining Eqs. (A5.4) and (A5.7) as

$$\beta = \frac{n(t,0)}{N(t)} = \left[\int_0^\infty l(x)e^{-\lambda x} dx \right]^{-1}. \qquad (A5.8)$$

If the asymptotic population growth rate can be decomposed into births and deaths as $\lambda = \beta - \delta$, where δ (day^{-1}) is the specific death rate of the stable age distribution, the total population size will be described by the differential equation

$$\dot{N}(t) = \lambda N(t) = (\beta - \delta)N(t). \qquad (A5.9)$$

Finally, the frequency distribution function, $f_n(x)$, of the stable age distribution can written as

$$f_n(x) = \frac{n(t,x)}{N(t)} = \beta l(x)e^{-\lambda x}. \qquad (A5.10)$$

In other words, the stable age distribution is proportional to the survival function depreciated by the intrinsic rate of increase, such that the fraction of old individuals in the population will decrease with increasing population growth rate. The specific birth rate will be a proportionality factor ensuring that Eq. (A5.10) is a proper distribution function (i.e., the cumulative distribution function corresponding to Eq. (A5.10) becomes equal to 1 as $x \to \infty$).

A6 Stationary Points of the Nutrients, Algae, and Herbivores Model

In the theory of dynamic systems (e.g., Luenberger 1979), a system is described by n state variables $x_1, \cdots, x_n$, which are related through a set of ordinary differential equations $\dot{x}_i = f_i(x_1, \cdots, x_n)$ for $i = 1, \cdots, n$. In more compact vector notation, this is written as $\dot{\mathbf{x}} = \mathbf{f}(\mathbf{x})$, where $\mathbf{x}^T = (x_1, \cdots, x_n)$ is the state vector and $\mathbf{f}^T = (f_1, \cdots, f_n)$ is a vector of functions in $\mathbf{x}$. A stationary point $\bar{\mathbf{x}}$ is defined as a point in the state space where all time derivatives vanish ($\dot{\mathbf{x}} = 0$). If a system is located at a stationary point, it will stay there indefinitely, unless it by some means is perturbed away from the stationary point. The stationary points are found by solving the set of equations $\mathbf{f}(\bar{\mathbf{x}}) = 0$; if the equations are nonlinear, there can be many of them or, in some cases, there might be no stationary points at all.

The set of equations determining the stationary points of Eqs. (5.1)-(5.3) are found by setting the time derivatives to zero:

$$DP_L - (\sigma + D)P - \theta gZ = 0 \qquad\qquad\qquad (A6.1)$$

$$(\mu - (\sigma + D))C - IZ = 0 \qquad\qquad\qquad (A6.2)$$

$$(g - (\delta + D))Z = 0 . \qquad\qquad\qquad (A6.3)$$

Three of the coefficients in Eqs. (A6.1)-(A6.3) are nonlinear functions (μ, I, g) of the state variables (P, C, Z), such that the steady-state equations will be nonlinear. Among the five stationary points that can be solutions of Eqs. (A6.1)-(A6.3), we will start by examining two apparently trivial ones, which are nevertheless important in determining the dynamics of the system Eqs. (5.1)-(5.3).

The Washout Point. This stationary point corresponds to the lifeless situation where both the biomasses of algae and grazers are zero ($C = 0$ and $Z = 0$). Due to the simplifying assumption of neglecting dissolved P, the concentration of algal P will still be nonzero:

$$P = \frac{D}{\sigma + D} P_L . \qquad\qquad\qquad (A6.4)$$

This apparent contradiction between zero algal C and nonzero algal P is resolved if the state variable P is interpreted as the amount of P potentially available for algal growth. Formally, the algal P quota will be infinite at this stationary point, but the algal growth rate will still be finite and equal to the asymptotic maximum ($\mu = \mu'$) because of the hyperbolic relationship between cell quota and growth rate [Eq. (3.2)].

The Grazer Extinction Point. This stationary point corresponds to the ungrazed situation where grazer biomass is zero ($Z = 0$), while algal biomass is nonzero ($C \neq 0$). From Eq. (A6.2), the conditions $Z = 0$ and $C \neq 0$ together imply that $\mu = \sigma + D$, or that algal growth rate is exactly balanced by losses from flushing and sedimentation. The algal P concentration (P) will still be determined by P supply rate, flushing, and sedimentation as in Eq. (A6.4), so that the algal biomass at this stationary point is found by substituting $\mu = \sigma + D$ in the Droop equation [Eq. (3.2)], solving for the steady-state cell quota Q, and combining this result with the requirement that $C = P/Q$:

$$C = \frac{\mu' - (\sigma + D)}{\mu' Q'} \frac{D}{\sigma + D} P_L. \tag{A6.5}$$

In other words; algal biomass will be directly proportional to the P supply rate and non-negative for all dilution rates $D \leq \mu' - \sigma$. At a critical dilution rate $D = \mu' - \sigma$, the grazer extinction point will be coincident with the washout point; the algae are unable to compensate for the losses through flushing and sedimentation, and positive net growth thus becomes impossible.

Apart from the special case where a grazer population growing at its innate capacity is exactly balanced by the losses through the outflow ($D = g' - \delta$), the grazer growth rate must be controlled by the properties of the prey population (so that $g = \delta + D$) in order to have a stationary point with nonzero grazer biomass ($Z \neq 0$). We can distinguish between three cases as grazer growth rate can be limited by either food composition, food concentration, or both. The use of piecewise linear relationships describing ingestion and utilization of food implies that the stationary points will be determined by a different set of equations for each of the three cases.

P-Limited Grazer Growth. This stationary point corresponds to the situation where algal P content is below the requirements of the grazer ($Q < \theta$), while algal food carbon is above the incipient limiting level and thus non-limiting to grazer growth ($C \geq C'$). This means that food ingestion is at its maximal value ($I = I'$) and Eq. (A6.3) can be written as

$$g = (\varepsilon I' - r)\frac{Q}{\theta} = \delta + D. \tag{A6.6}$$

By solving Eq. (A6.6) for algal P content (Q), it is seen that Q will be a constant which depends on the dilution rate, but not on the input P concentration (P_L). Thus, algal growth rate will also be a constant which will be denoted by

$$\bar{\mu} = \mu'\left(1 - \frac{g'}{\delta + D} \frac{Q'}{\theta}\right), \tag{A6.7}$$

where $g' = \varepsilon I' - r$ is the maximal grazer growth rate. By eliminating Z from Eqs. (A6.1) and (A6.2), utilizing the relationship between P and C implied by Eq. (A6.6), the state variables at the stationary point can be expressed recursively as

$$P = \frac{D I'}{(\sigma + D)I' + (\bar{\mu} - (\sigma + D))g'} P_L, \tag{A6.8}$$

$$C = \frac{\delta + D}{g'} \frac{P}{\theta}, \text{ and} \tag{A6.9}$$

$$Z = \frac{\bar{\mu} - (\sigma + D)}{I'} C.$$ (A6.10)

By examining the stationary state given by Eqs. (A6.8)-(A6.10), it is seen that as long as net algal growth is possible ($\bar{\mu} = \delta + D$), both algal and grazer biomasses will be positively correlated and will increase in proportion to the input P concentration (P_L). The condition for existence of this stationary point will be that grazer biomass is strictly positive, which is equivalent to requiring that $\bar{\mu} > \sigma + D$. If $\bar{\mu} = \delta + D$, then the stationary point will be identical to the grazer extinction point discussed above.

Simultaneously C- and P-Limited Grazer Growth. This stationary point corresponds to the situation where both algal P content is below the requirements of the grazer ($Q < \theta$), and algal food carbon is below the incipient limiting level ($C < C'$). Food carbon ingestion is then proportional to algal biomass, such that after substituting $I = I'(C/C')$, Eq. (A6.2) can be written as

$$[(\mu - (\sigma + D)) - (I'/C')Z]C = 0.$$ (A6.11)

As $C \neq 0$, the steady-state zooplankton biomass at a given algal growth rate (μ) will be

$$Z = \frac{\mu - (\sigma + D)}{I'} C'.$$ (A6.12)

Grazer biomass will be non-negative for all $\mu \geq \sigma + D$; for $\mu = \sigma + D$ this stationary point will be identical to the grazer extinction point ($Z = 0$). P-limited grazer growth requires that $Q < \theta$, which again implies that, $\mu < \mu_\theta$ with μ_θ defined as

$$\mu_\theta = \mu'\left(1 - \frac{Q'}{\theta}\right).$$ (A6.13)

If we define

$$Z_\theta = \frac{\mu_\theta - (\sigma + D)}{I'} C',$$ (A6.14)

then zooplankton biomass will be constrained to the interval $0 < Z < Z_\theta$ whenever $\sigma + D < \mu < \mu_\theta$. The linear relationship between algal growth rate and zooplankton biomass implies that algal P content might equally well be expressed in terms of Z. If we substitute the Droop equation [Eq. (3.2)] into Eq. (A6.12) and solve for Q, we obtain

$$\frac{Q}{\theta} = \frac{Z' - Z_\theta}{Z' - Z},$$ (A6.15)

where Z' is defined in analogy with Z_θ as

$$Z' = \frac{\mu' - (\sigma + D)}{I'} C'.$$

(A6.16)

When grazer biomass is nonzero, Eq. (A6.3) implies that $g = \delta + D$, or that

$$(\varepsilon I'(C/C') - r)(Q/\theta) = \delta + D.$$

(A6.17)

If we substitute Eq. (A6.15) into (A6.17) and solve for C, we find that steady-state algal biomass can be expressed in terms of grazer biomass as

$$C = C_0'' + \frac{Z' - Z}{Z' - Z_\theta} (C'' - C_0'').$$

(A6.18)

Algal biomass will thus be linearly decreasing with increasing grazer biomass (and increasing algal growth rate), such that $C = C''$ when $Z = Z_\theta$. For $\mu = \sigma + D$ (or $Z = 0$), C will be equal to the algal biomass at the grazer extinction point [Eq. (A6.5)]. The parameter C'' is defined as

$$C'' = \frac{r + \delta + D}{\varepsilon I'} C',$$

(A6.19)

and can be interpreted as a threshold food concentration for positive net growth in the grazer population when food composition is optimal; that is, $g = \delta + D$ when $C = C''$ and $Q \geq \theta$. The other parameter C_0'', which is defined as

$$C_0'' = \frac{r}{\varepsilon I'} C',$$

(A6.20)

can be interpreted as a threshold food concentration for positive gross grazer growth rate on optimal food; that is, $g = 0$ when $C = C_0''$ and $Q \geq \theta$. Steady-state algal biomass will be below the incipient limiting level ($C < C''$) as long as $\varepsilon I' > r$, which must be true in all cases where the grazers are capable of positive net growth.

From the relationship $P = QC$ and Eqs. (A6.15) and (A6.18), the concentration of algal P at the stationary point can be expressed as

$$P = \theta \left(C'' - \frac{Z_\theta - Z}{Z' - Z} C_0'' \right),$$

(A6.21)

so that P will be an increasing function of Z, with $P = \theta C''$ at $Z = Z_\theta$.

Thus far, the location of the stationary point has been expressed in terms of the specific algal growth rate (μ), or the linearly related steady state grazer biomass (Z). By substituting Eq. (A6.21) into (A6.1), we could have obtained a quadratic equation in Z which could have been solved to give an explicit expression of Z [and also C and P, through Eqs. (A6.18) and (A6.21)] in terms of the input P concentration P_L. For spreadsheet calculations and graphical display, it is more efficient to choose a value of μ

within the range $\sigma + D < \mu < \mu_\theta$ and calculate the steady-state values of Z, C, and P from Eqs. (A6.12), (A6.18), (A6.21). The corresponding input P concentration can then be found from solving Eq. (A6.1) as

$$P_L = \left(1 + \frac{\sigma}{D}\right)P + \left(1 + \frac{\delta}{D}\right)\theta Z. \tag{A6.22}$$

By substituting Eq. (A6.21) into (A6.22), it is seen that $Z > 0$ implies that the input P concentration must be above a critical level ($P_L > P'_L$), with P'_L given by

$$P'_L = \left(1 + \frac{\sigma}{D}\right)\left(C'' - \frac{Z_\theta}{Z'}C_0''\right)\theta. \tag{A6.23}$$

By the same kind of reasoning, we can see that $Z \leq Z_\theta$ implies that the input P concentration must be below a critical level ($P'_L \leq P''_L$), with P''_L given by

$$P''_L = \left(\left(1 + \frac{\sigma}{D}\right)C'' + \left(1 + \frac{\delta}{D}\right)Z_\theta\right)\theta. \tag{A6.24}$$

The domain of the stationary point with both C- and P-limited grazer growth will therefore be given in terms of the P loading conditions as $P'_L < P_L \leq P''_L$. When the input P concentration increases from P'_L to P''_L, steady-state algal growth rate will also increase from $\sigma + D$ to μ_θ. Over the same range of P loading conditions, steady state grazer biomass will increase from 0 to Z_θ, while algal biomass will decrease to C''.

C-Limited Grazer Growth. This stationary point corresponds to the classical situation of food-limited grazer growth where algal P content is sufficient for the requirements of the grazer ($Q \geq \theta$), while algal food carbon is below the incipient limiting level ($C < C'$). This case also represents the direct continuation of the previous case to the situation where $Q \geq \theta$.

The requirement $Q \geq \theta$, means that specific algal growth rate (μ) will be constrained to $\mu_\theta \leq \mu \leq \mu'$ over the domain of this stationary point. As $C < C'$, Eq. (A6.2) will also for this stationary point yield a linear relationship, identical to Eq. (A6.12), between μ and Z, implying that grazer biomass will be constrained to $Z_\theta \leq Z \leq Z'$. The absence of a food composition effect on grazer growth rate when $Q \geq \theta$ implies that under the assumption that $Z \neq 0$, Eq. (A6.3) can be written as

$$\varepsilon I'(C/C') - r = \delta + D. \tag{A6.25}$$

Solving Eq. (A6.25) for C gives that the steady state algal biomass is constant $C = C''$ over the whole domain of this stationary point. From the relationship $P = QC$ and Eq. (A6.15), the concentration of algal P at the stationary point can be expressed as

$$P = \frac{Z' - Z_\theta}{Z' - Z}\theta C'', \tag{A6.26}$$

so that P will be an increasing function of Z, with $P = \theta C''$ at $Z = Z_\theta$. The singularity at $Z = Z'$ implies that $P \to \infty$ when $Z \to Z''$, or, by reversing the argument, that grazer biomass increases asymptotically to Z' as $P \to \infty$. As Z remains finite, Eq. (A6.22) implies that $P \to \infty$ as $P_L \to \infty$. The domain of this stationary point in terms of input P concentration will thus be $P_L \geq P''_L$. Across the boundary of this domain, the trajectory of the stationary point with C-limited grazer growth will be continuous with the trajectory of the stationary point of simultaneous C- and P-limitation.

A7 Local Stability Analysis of the Nutrients, Algae, and Herbivores Model

Consider a general non-linear dynamic system $\dot{x} = f(x)$, with a stationary point $\bar{x}$ such that $f(\bar{x}) = 0$. If the system is perturbed a small distance δx from the stationary point, such that $x = \bar{x} + \delta x$, the effect of this perturbation can be studied by considering the locally linearized system $\dot{\delta x} = A\delta x$, where A is the Jacobian matrix of the non-linear system, evaluated at the stationary point (for further details, see Luenberger 1979). The Jacobian is defined as the matrix $A = \partial f / \partial x^T$ of partial derivatives with respect to the state variables; that is, an element $a_{i,j}$ of A is the partial derivative of equation i by state variable j ($a_{i,j} = \partial f_i / \partial x_j$).

The behavior of the linear approximation to a small perturbation will be determined by the signs of the eigenvalues of the Jacobian matrix. If the real parts of all the eigenvalues of the Jacobian are negative, then small perturbations will decay exponentially with time such that the system will eventually return to the stationary point. A system with this behavior is said to be locally, asymptotically stable. If any eigenvalue has a positive real part, then any small perturbation from the equilibrium can grow exponentially with time, leading eventually to the system being driven away from the stationary point. If the functions $f(x)$ are analytical everywhere in the state space (i.e., have continuous derivatives with respect to x), then the existence of a single, locally stable stationary point implies that this point will be a global equilibrium, where the system will end up in from any arbitrary initial condition. In more general cases, with multiple stationary points having different local stability properties, the relationship between local and global stability can be very complex (Guckenheimer and Holmes 1983).

The eigenvalues $\lambda_1, \cdots, \lambda_n$ of a general $n \times n$ matrix A are found by solving the characteristic equation $|\lambda I - A| = 0$, which is equivalent to solving an n'th order polynomial equation in λ. Fortunately, for the case of local stability analysis, some properties of the roots of a general polynomial equation can be investigated without explicitly solving the equation (which is possible in only very special cases for $n > 3$, and usually leads to very complex expressions, even for $n > 2$). The *Routh-Hurwitz criterion* (see Luenberger 1979) gives information on the signs of the roots of a polynomial just from relations between the coefficients of the polynomial. For the third-order equation that will appear repeatedly in the forthcoming sections, the Routh-Hurwitz criterion can be stated as: the roots of the characteristic polynomial $\lambda^3 + c_2\lambda^2 + c_1\lambda + c_0 = 0$ will all have negative real parts, if the coefficients are such that $c_0 > 0$, $c_2 > 0$, and $c_0c_2 - c_0 > 0$.

Taking the partial derivatives with respect to the state variables (P, C, and Z) of the state equations (5.1)-(5.3), gives a Jacobian matrix that can be written as

$$
A = \begin{bmatrix}
-(\sigma + D) - \theta\dfrac{\partial g}{\partial P}Z & -\theta\dfrac{\partial g}{\partial C}Z & -\theta g \\[2mm]
\dfrac{\partial \mu}{\partial P}C & \dfrac{\partial \mu}{\partial C}C + \mu - (\sigma + D) - \dfrac{\partial I}{\partial C}Z & -I \\[2mm]
\dfrac{\partial g}{\partial P}Z & \dfrac{\partial g}{\partial C}Z & g - (\delta + D)
\end{bmatrix} \qquad \text{(A7.1)}
$$

The Jacobian matrix as it stands in Eq. (A7.1) reveals little about the properties of the system. More insight can be gained from taking advantage of the simplifications that are possible when investigating the local stability properties of the individual stationary points that were identified in Appendix A6.

The Washout Point. This stationary point has the property that both algal and grazer biomasses are zero ($C = 0$ and $Z = 0$), which means that all terms involving C or Z in Eq. (A7.1) will also be zero. P being non-zero, while C is zero, means that the algal P content (Q) will formally be infinite. This implies that algal growth rate will be at the asymptotic maximum ($\mu = \mu'$), and that grazer growth rate (g) will be independent of food P content ($Q > \theta$). As no food means no ingestion ($I = 0$), the grazer growth rate will be equal to respiration losses ($g = -r$).

$$
A = \begin{bmatrix}
-(\sigma + D) & 0 & \theta r \\
0 & \mu' - (\sigma + D) & 0 \\
0 & 0 & -(r + \delta + D)
\end{bmatrix}. \qquad \text{(A7.2)}
$$

From the particularly simple Jacobian matrix at this stationary point [Eq. (A7.2)], the characteristic equation ($|\lambda I - A| = 0$) can readily be found:

$$
(\lambda + (\sigma + D))(\lambda - (\mu' - (\sigma + D)))(\lambda + (r + \delta + D)) = 0 \qquad \text{(A7.3)}
$$

[in fact, the eigenvalues will simply be the diagonal elements in an upper triangular matrix, like Eq. (A7.2)]. It can be see from Eq. (A7.3) that under the condition $D < \mu' - \sigma$, this stationary point will have two negative and one positive eigenvalues [$-(\sigma + D)$, $-(r + \delta + D)$], and [$\mu' - (\sigma + D)$], and therefore be locally unstable. On the other hand, when $D > \mu' - \sigma$, all eigenvalues will be negative, so that the washout condition will be locally stable, as can be expected. This means that a small inoculum of algae and grazers will be able to invade a sterile system, unless the dilution rate is so high that the algae are unable to balance the losses through the outflow.

The Grazer Extinction Point. This stationary point has the property that grazer biomass is zero ($Z = 0$), which means that all terms involving Z in Eq. (A7.1) will also be zero. Since algal biomass is non-zero ($C \neq 0$), Eq. (A6.2) implies that algal growth must be equal to dilution and sedimentation losses ($\mu = \sigma + D$).

Nonzero algal biomass (C) means that we cannot eliminate the partial derivatives of μ as in the previous case. If the Droop equation [Eq. (3.2)] is written as

$$\mu = \mu'\left(1 - Q'\frac{C}{P}\right),\qquad\qquad\qquad\text{(A7.4)}$$

then the terms in Eq. (A7.1) involving the partial derivatives of μ with respect to P and C can be expressed as

$$\frac{\partial \mu}{\partial P}C = \mu'Q'\left(\frac{C}{P}\right)^2 = \frac{1}{\mu'Q'}\left(\mu'Q'\frac{C}{P}\right)^2 = \frac{(\mu'-\mu)^2}{\mu'Q'}\qquad\text{(A7.5)}$$

$$\frac{\partial \mu}{\partial C}C = -\mu'Q'\frac{C}{P} = -(\mu'-\mu).\qquad\qquad\text{(A7.6)}$$

The right most identities in Eqs. (A7.5), (A7.6) comes from solving Eq. (A7.4) for $\mu' - \mu$. Substituting $\mu = \sigma + D$ into Eqs. (A7.5) and (A7.6) yields the Jacobian matrix for the grazer extinction point:

$$\mathbf{A} = \begin{bmatrix} -(\sigma+D) & 0 & -\theta g \\ (\mu'-(\sigma+D))^2/(\mu'Q') & -(\mu'-(\sigma+D)) & -I \\ 0 & 0 & g-(\delta+D) \end{bmatrix}. \text{(A7.7)}$$

Although the Jacobian matrix [Eq. (A7.7)] contains more non-zero entries than Eq. (A7.2), the characteristic polynomial corresponding to $|\lambda I - A| = 0$ will still have the same simple structure as Eq. (A7.3):

$$(\lambda + (\sigma+D))(\lambda + (\mu'-(\sigma+D)))(\lambda - (g-(\delta+D))) = 0. \quad \text{(A7.8)}$$

When the dilution rate is below the washout rate ($D < \mu' - \sigma$), two of the roots of Eq. (A7.8) will be negative [$-(\sigma+D)$ and $-(\mu'-(\sigma+D))$], such that the local stability of the extinction point will depend on the sign of $g - (\sigma + D)$. If the grazers are able to have positive, net population growth on algae growing at $\mu = \sigma + D$, the grazer extinction point will be locally unstable; if not, it will be locally stable. In other words, the grazers will be able to invade an algal community at equilibrium with dilution and sedimentation losses only if they are able to maintain net population growth on this food resource.

P-Limited Grazer Growth. In Appendix A6 it was shown that this stationary point has the property that both algal and grazer biomasses are nonzero ($C > 0$ and $Z > 0$) and proportional to the input P concentration (P_L), while the algal growth rate (μ) is a constant given by Eq. (A6.7), independent of P_L. It was further shown that the existence of this stationary point (that is, with $Z > 0$) requires that $\mu > \sigma + D$.

In contrast to the previous two cases, the presence of a nonzero grazer population means that terms in Eq. (A7.1) involving partial derivatives of g and I must be evaluated. As food carbon concentration is by definition nonlimiting to grazer growth at this stationary point, we must have that $C \geq C'$ and thus $I = I'$, which means that the partial derivative of I with respect to C is zero. If we express the grazer growth rate as

$$g = (\varepsilon I' - r)\frac{Q}{\theta} = g'\frac{Q}{\theta} = g'\frac{P}{\theta C}, \qquad (A7.9)$$

then the term involving partial derivatives of g by P in Eq. (A7.1) becomes

$$\frac{\partial g}{\partial P}Z = \frac{g'}{\theta}\frac{Z}{C} = \frac{g'}{\theta}\frac{\mu - (\sigma + D)}{I'} = K'(\mu - (\sigma + D))\theta^{-1}. \qquad (A7.10)$$

The second equality in Eq. (A7.10) comes from the constant ratio between grazer and algal biomass implied by Eq. (A6.10), while the last equality is from introducing the maximal gross growth efficiency $K' = g''/I'$. By the same reasoning, the term involving partial derivatives of g by C in Eq. (A7.1) becomes

$$\frac{\partial g}{\partial C}Z = -\frac{g'}{\theta}\frac{P}{C}\frac{Z}{C} = -K'(\mu - (\sigma + D))\frac{Q}{\theta} \qquad (A7.11)$$

$$= -K'(\mu - (\sigma + D))\frac{\mu' - \mu_\theta}{\mu' - \mu}$$

The last equality in Eq. (A7.11) comes from utilizing the relationship between algal P content and growth rate implied by the Droop equation [Eq. (3.2)]:

$$\frac{Q}{\theta} = \frac{\dfrac{\mu'Q'}{\mu' - \mu}}{\dfrac{\mu'Q'}{\mu' - \mu_\theta}} = \frac{\mu' - \mu_\theta}{\mu' - \mu}, \qquad (A7.12)$$

and introducing the growth rate μ_θ corresponding to $Q = \theta$, as defined by Eq. (A6.13). Using Eq. (A6.13), we can reexpress Eq. (A7.5) such that the terms involving partial derivatives of μ can be written as

$$\frac{\partial \mu}{\partial P}C = \frac{(\mu' - \mu)^2}{\mu' - \mu_\theta}\theta^{-1} \text{ and} \qquad (A7.13)$$

$$\frac{\partial \mu}{\partial C} C = -(\mu' - \mu).$$ (A7.14)

By substitution of Eq. (A7.12) into (A7.9), we can also reexpress the grazer growth rate at the stationary point as

$$g = K'I' \frac{\mu' - \mu_\theta}{\mu' - \mu}.$$ (A7.15)

The Jacobian of the stationary point with P-limited grazer growth is found by substituting Eqs. (A7.10), (A7.11), (A7.13), (A7.14), (A7.15) into (A7.1), and noticing that Eq. (A6.3) implies that $g - (\delta + D) = 0$ if $Z > 0$. If we introduce the parameters

$$\omega_1 = \mu - (\sigma + D),$$ (A7.16)

$$\omega_2 = \mu' - \mu, \text{ and}$$ (A7.17)

$$\omega_3 = \mu' - \mu_\theta,$$ (A7.18)

which will all be positive quantities as long as the stationary point exists (that is, when $\sigma + D < \mu < \mu'$), then the Jacobian can be written concisely as

$$\mathbf{A} = \begin{bmatrix} -(\sigma + D) - K'\omega_1 & K'\omega_1\omega_2^{-1}\omega_3\theta & -K'I'\omega_2^{-1}\omega_3\theta \\ \omega_2^2\omega_3^{-1}\theta^{-1} & \omega_1 - \omega_2 & -I' \\ K'\omega_1\theta^{-1} & -K'\omega_1\omega_2^{-1}\omega_3 & 0 \end{bmatrix}.$$ (A7.19)

The characteristic equation ($|\lambda \mathbf{I} - \mathbf{A}| = 0$) of Eq. (A7.19) will be a third-order polynomial equation $\lambda^3 + c_2\lambda^2 + c_1\lambda + c_0 = 0$, with coefficients

$$c_2 = -(1 - K')\omega_1 + \omega_2 + (\sigma + D),$$ (A7.20)

$$c_1 = (\sigma + D)(\omega_2 - \omega_1) - \omega_1 \left(K'\omega_1 + (1 - K')I'\omega_2^{-1}\omega_3\right), \text{ and}$$ (A7.21)

$$c_0 = -K'I'\omega_1\left(K'\omega_1 + (\sigma + D)\right)\omega_2^{-1}\omega_3$$ (A7.22)

By inspecting Eq. (A7.22) it is seen that since all terms in the expression are positive, the coefficient c_0 must be negative. In other words, the Routh-Hurwitz criterion ($c_0 > 0$, $c_2 > 0$, and $c_1c_2 - c_0 > 0$) cannot be satisfied for this stationary point, so it cannot be locally stable.

Simultaneously C- and P-Limited Grazer Growth. In Appendix A6 it was shown that this stationary point exists only for input P concentration in the range $P'_L < P_L \le P''_L$. As P_L increases from P'_L to P''_L, algal growth rate increases from $\sigma - D$ to μ_q [defined by Eq. (A6.13)]. Over the same range, grazer biomass increases from zero to Z_θ, given by Eq. (A6.14), while algal biomass decreases to the threshold level for net grazer growth on optimal food, C''.

As algal biomass is below the incipient limiting level ($C < C'$), the ingestion rate will be a function of C [$I = I'(C/C')$], and thus the term involving the partial derivative of I with respect to C will be nonzero:

$$\frac{\partial I}{\partial C} Z = \frac{I'}{C'} Z = \mu - (\sigma + D) \cdot \qquad (A7.23)$$

The second identity in Eq. (A7.23) comes from observing that when $C \neq 0$, the equilibrium condition [Eq. (A6.2)] becomes $(\mu - (\sigma + D))C' = I'Z$.

When both algal P content is below the requirements of the grazer ($Q < \theta$), and algal food carbon is below the incipient limiting level ($C < C'$), the grazer growth rate (g) becomes

$$g = \left(\varepsilon I' \frac{C}{C'} - r \right) \frac{Q}{\theta} = \left(\varepsilon I' \frac{C}{C'} - r \right) \frac{P}{\theta C} \cdot \qquad (A7.24)$$

The Jacobian terms involving the partial derivative of g with respect to P are then

$$\frac{\partial g}{\partial P} Z = (\varepsilon I - r) \frac{1}{\theta C} Z = \frac{\varepsilon I - r}{I}(\mu - (\sigma + D))\theta^{-1} \qquad (A7.25)$$

$$= K_P(\mu - (\sigma + D))\theta^{-1}$$

The second equality in Eq. (A7.25) comes from rearranging Eq. (A6.2) and substituting the ratio between grazers and algae: $Z/C = (\mu - (\sigma + D))/I$. The last identity in Eq. (A7.25) comes from introducing the dimensionless gross phosphorus growth efficiency

$$K_P = \frac{\theta g}{QI} = \frac{\varepsilon I - r}{I}, \qquad (A7.26)$$

which can be interpreted as the quantity of zooplankton P produced per unit of algal P ingested. For $Z > 0$, the equilibrium condition [Eq. (A6.3)] implies $\gamma = \delta + D > 0$, which again implies that $I > \varepsilon I > r$, or $\varepsilon > K_P > 0$. By the same kind of argument, the Jacobian terms involving the partial derivative of g with respect to C can be expressed as

$$\frac{\partial g}{\partial C} Z = -r \frac{P-1}{\theta C^2} Z = r \frac{Q}{\theta} \frac{Z}{C} \qquad (A7.27)$$

$$= \frac{r}{I}(\mu - (\sigma + D)) \frac{Q}{\theta} = (\varepsilon - K_P)(\mu - (\sigma + D)) \frac{Q}{\theta}$$

$$= (\varepsilon - K_P)(\mu - (\sigma + D)) \frac{\mu' - \mu_\theta}{\mu' - \mu}$$

The third identity in Eq. (A7.27) is from the same use of Eq. (A6.2) as in Eq. (A7.25), while the fourth identity is from observing that $r/I = \varepsilon - K_P$, and the last identity from the relationship between algal P content and growth rate as expressed in Eq. (A7.12). By rearranging Eq. (A7.26), we can express g in terms of I and K_P as

$$g = K_P I \frac{Q}{\theta} = K_P I \frac{\mu' - \mu_\theta}{\mu' - \mu}. \tag{A7.28}$$

As terms involving partial derivatives of μ will be the same as in the previous case, the Jacobian of this stationary point is found by substituting Eqs. (A7.13), (A7.14), (A7.23), (A7.25), (A7.27), (A7.28) into Eq. (A7.1). If we make use of the set of positive parameters ω_1, ω_2, and ω_3 as defined by Eqs. (A7.16)-(A7.18), the Jacobian can be written as

$$\mathbf{A} = \begin{bmatrix} -((\sigma+D)+K_P\omega_1) & -(\varepsilon - K_P)\omega_1\omega_2^{-1}\omega_3\theta & -K_P I \omega_2^{-1}\omega_3\theta \\ \omega_2^2\omega_3^{-1}\theta^{-1} & -\omega_2 & -I \\ K_P\omega_1\theta^{-1} & (\varepsilon - K_P)\omega_1\omega_2^{-1}\omega_3 & 0 \end{bmatrix}. \tag{A7.29}$$

The characteristic equation ($|\lambda I - A| = 0$) of Eq. (A7.29) will be a third-order polynomial equation $\lambda^3 + c_2\lambda^2 + c_1\lambda + c_0 = 0$, with coefficients

$$c_2 = (\sigma + D) + K_P\omega_1 + \omega_2, \tag{A7.30}$$

$$c_1 = (\varepsilon\omega_1 + (\sigma+D))\omega_2 + (\varepsilon - (1 - K_P)K_P)I\omega_1\omega_2^{-1}\omega_3, \text{ and} \tag{A7.31}$$

$$c_0 = (\varepsilon K_P\omega_2 + (\varepsilon - K_P)(\sigma+D))I\omega_1\omega_2^{-1}\omega_3. \tag{A7.32}$$

As $1 > \varepsilon > K_P > 0$, we must have $\varepsilon(1 - K_P)K_P > \varepsilon - K_P > 0$; this means that all terms of c_2, c_1, and c_0 are positive, and that the first two Routh-Hurwitz criteria ($c_0 > 0$ and $c_2 > 0$) will be satisfied. The last criterion ($c_1c_2 - c_0 > 0$) can, after some manipulation, be written as

$$c_1c_2 - c_0 = (K_P\omega_1 + \omega_2 + (\sigma+D))(\varepsilon\omega_1 + (\sigma+D))\omega_2$$
$$+ K_P(\varepsilon - (1 - K_P)K_P)I\omega_1^2\omega_2^{-1}\omega_3 \tag{A7.33}$$
$$+ (1 - K_P)(\varepsilon - K_P)I\omega_1\omega_3$$
$$+ K_P^2(\sigma+D)I\omega_1\omega_2^{-1}\omega_3$$

and will also be positive under the condition that $1 > \varepsilon > \varepsilon - (1 - K_P)K_P > \varepsilon - K_P > 0$. In other words, all eigenvalues must have negative real parts, and the stationary point with simultaneous C- and P-limited grazer growth will therefore be a locally stable equilibrium.

C-Limited Grazer Growth. In Appendix A6 it was shown that this stationary point represents the continuation of the previous one for input P concentration $P_L \geq P''_L$. As P_L increases from P''_L towards infinity, algal growth rate increases from μ_q [defined by Eq. (A6.13)] to the asymptotic value μ'. Over the same range, grazer biomass increases from Z_θ, given by Eq. (A6.14), to an asymptotic level Z' given by Eq. (A6.16). Algal biomass is constant at the threshold level for net grazer growth on optimal food (C''), at this stationary point.

The terms in Eq. (A7.1) involving partial derivatives of μ with respect to C and P can be expressed identically to the previous case by Eqs. (A7.13) and (A7.14). As the ingestion rate (I) is still controlled by algal biomass (C), like in the previous case, the term involving the partial derivative of I by C can be expressed by Eq. (A7.23) in this case also. Since grazer growth is independent of algal P content when $Q > \theta$, all terms involving partial derivatives of g with respect to P vanish at this stationary point. The grazer growth rate at this stationary point is expressed as

$$g = \varepsilon I' \frac{C}{C'} - r, \tag{A7.34}$$

so that the term involving partial derivatives of g with respect to C becomes

$$\frac{\partial g}{\partial C} Z = \varepsilon \frac{I'}{C'} Z = \varepsilon(\mu - (\sigma + D)). \tag{A7.35}$$

The second identity in Eq. (A7.35) comes from the same utilization of the equilibrium condition [Eq. (A6.2)] as in Eqs. (A7.25) and (A7.27). By noticing that at this stationary point $g = \varepsilon I - r = \delta + D$, we can reexpress both g and I in terms of the constant parameters $\varepsilon, r, \delta,$ and D:

$$g = \delta + D. \tag{A7.36}$$

$$I = \varepsilon^{-1}(r + \delta + D) \tag{A7.37}$$

The Jacobian of this stationary point is found by substituting Eqs. (A7.13), (A7.14), (A7.35), (A7.36), (A7.37) into Eq. (A7.1). If we make use of the set of non-negative parameters $\omega_1, \omega_2,$ and ω_3 (ω_2 will be zero for $\mu = \mu'$), defined by Eqs. (A7.16)-(A7.18), the Jacobian can be written as

$$\mathbf{A} = \begin{bmatrix} -(\sigma + D) & -\varepsilon\omega_1\theta & -(\delta + D)\theta \\ \omega_2^2\omega_3^{-1}\theta^{-1} & -\omega_2 & -\varepsilon^{-1}(r + (\delta + D)) \\ 0 & \varepsilon\omega_1 & 0 \end{bmatrix}. \tag{A7.38}$$

The characteristic equation ($|\lambda \mathbf{I} - \mathbf{A}| = 0$) of Eq. (A7.38) will be a third-order polynomial equation $\lambda^3 + c_2\lambda^2 + c_1\lambda + c_0 = 0$, with coefficients

$$c_2 = \omega_2 + (\sigma + D), \tag{A7.39}$$

$$c_1 = (\sigma + D)\omega_2 + (r + (\delta + D) + \varepsilon\omega_2^2\omega_3^{-1})\omega_1, \text{ and} \tag{A7.40}$$

$$c_0 = ((\sigma + D)(r + (\delta + D)) + \varepsilon(\delta + D)\omega_2^2\omega_3^{-1})\omega_1. \tag{A7.41}$$

Since all the coefficients of the characteristic polynomial will be positive, the two first Routh-Hurwitz criteria are fulfilled. The last Routh-Hurwitz criterion becomes

$$c_1 c_2 - c_0 = \left(r\omega_3 + (\delta + D)(\omega_3 - \varepsilon\omega_2)\right)\omega_1$$
$$+ \left((\sigma + D) + \omega_2\right)\left(\varepsilon\omega_1\omega_2 + (\sigma + D)\omega_3\right) \tag{A7.42}$$

The term $\omega_3 - \varepsilon\omega_2 > \omega_3 - \omega_2 = \mu - \mu_q$ in Eq. (A7.42) will be nonnegative as long as $\mu \geq \mu_q$, which is satisfied since it is a condition for existence of this stationary point. In other words: as all Routh-Hurwitz criteria are satisfied, all eigenvalues will have negative real parts, which again implies that the stationary point with C-limited grazer growth will be locally stable.

A8 The Persistence Boundary

In Appendix 7 it was shown that the existence of a locally stable stationary point with zero grazer biomass, called the grazer extinction point, will be possible only when the grazers are unable to maintain positive net growth rate $[g < (\delta + D)]$ on algae growing at equilibrium with their dilution and sedimentation losses $[\mu = (\sigma + D)]$.

If we assume that $Q < \theta$ and $C \geq C'$ at the grazer extinction point, then the grazer ingestion rate will be saturated $(I = I')$ and the grazer growth rate will thus be determined by algal P content alone. From the Droop equation [Eq. (3.2)], the phosphorus cell quota of algae growing at $\mu = (\sigma + D)$ will be

$$Q = \frac{\mu'}{\mu' - (\sigma + D)} Q', \tag{A8.1}$$

giving a grazer growth rate

$$g = g'\frac{Q}{\theta} = g'\frac{\mu'}{\mu' - (\sigma + D)}\frac{Q'}{\theta} = \frac{\mu' g''}{\mu' - (\sigma + D)}, \tag{A8.2}$$

where $g' = \varepsilon I' - r$ and $g'' = g'Q'/\theta$. The inequality $g < (\delta + D)$, determining the local stability of the grazer extinction point, can then be written as

$$(\delta + D)\left((\mu' - \sigma) - D\right) > \mu' g''. \tag{A8.3}$$

The implications of this quadratic inequality in D become somewhat clearer on looking at a simple illustration (Fig. A8.1). When viewed as functions of D, the left-hand side of Eq. (A8.3) is a concave parabola which intersects the D axis at $D = \mu' - \sigma$ and $D = -\delta$, and which is equal to $\delta(\mu' - \sigma)$ when $D = 0$, while the right-hand side is a straight line, parallel to the D axis, at distance $\mu' g''$. The local stability condition [Eq. (A8.3)] for the grazer extinction point is satisfied for the closed interval on the D axis where the parabola lies above the straight line. The bounds of this interval are determined by the roots of the quadratic equation $(\delta + D)((\mu' - \sigma) - D) = \mu' g''$, which can, after some manipulation, be expressed as

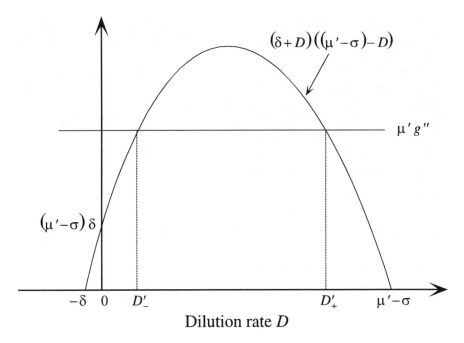

Fig. A8.1. Left- and right-hand sides of the inequality [Eq. (A6.3)], the local stability condition for the grazer extinction point, expressed as functions of the dilution rate D (see text for details)

$$D = \tfrac{1}{2}\left((\mu' - \sigma) - \delta\right) \pm \sqrt{\left(\tfrac{1}{2}\left((\mu' - \sigma) + \delta\right)\right)^2 - \mu' g''}\, . \qquad \text{(A8.4)}$$

If we denote the smallest and largest root of Eq. (A8.4) by D'_- and D'_+, the grazer extinction point will be locally stable for dilution rates satisfying $D'_- < D < D'_+$. If we define system persistence, sensu Gard and Hallam (1979), as the negation of grazer extinction, the system will be persistent for dilution rates $0 < D < D'_-$ or $D'_+ < D < \mu' - \sigma$.

By examining Fig. A8.1 it is seen that the persistence boundary at low dilution rates, D'_-, will be displaced to the left as g'' decreases. When $\mu' g'' \le (\mu' - \sigma)\delta$, or equivalently $g'' \le ((\mu' - \sigma)/\mu')\delta$, system persistence will be impossible at low dilution rates. If $\mu' >> \sigma$, the persistence condition will be approximately equal to $g'' > \delta$. In other words, grazers that are unable to balance their mortality losses when feeding on non-growing algae will eventually become extinct. When g'' is increased, the interval where the grazer extinction point is locally stable will decrease. In the limiting case where the apex of the parabola is below the line parallel to the D axis at distance $\mu'g''$, the grazer extinction point will be locally unstable, implying system persistence for all dilution rates $D < \mu' - \sigma$.

Grazer washout will result if the grazers are unable to compensate for dilution losses, even when growing at their maximum capacity; that is, when $D > g' - \delta$. In order to have a persistent system at high dilution rates, the requirements $D'_+ < D$ and $D < g' - \delta$ must therefore both be satisfied. The existence of any persistent system at high dilution rates thus requires that $D'_+ \leq g' - \delta$. We can proceed by observing that $\mu'g''$ can be expressed in terms of the algal growth rate at $Q = \theta\,[\mu_\theta$, given by Eq. (A6.13)] as

$$\mu'g'' = g'\,\mu'\,Q'/\theta = g'\,(\mu' - \mu_\theta).\qquad\qquad(A8.5)$$

If we substitute Eq. (A8.5) into (A8.4), we find that the requirement $D'_+ \leq g' - \delta$ is equivalent to $g' - \delta \geq \mu_\theta - \sigma$. This means that when the phosphorus contents of algae and grazers are equal, the grazers must have a maximum growth rate that is high enough to outgrow their prey in order to have a persistent system at a dilution rate $D > D'_+$.

We can gain more insight into how parameter changes will affect the persistence boundary D'_- by evaluating the partial derivatives of D'_- with respect to the parameters. As the parameters involved in Eq. (A8.4) can differ widely in magnitude, it is reasonable to discuss at relative parameter sensitivities, as introduced in Section 4.7 [the relative sensitivity, or elasticity, of a function z with respect to a parameter p is defined as $(p/z)\partial z/\partial p = \partial(\ln z)/\partial(\ln p)$] In order to simplify the expressions, we can define the dimensionless parameter ξ as

$$\xi = \left(1 - \frac{\delta + D'_-}{(\mu' - \sigma) - D'_-}\right)^{-1}.\qquad\qquad(A8.6)$$

The relative sensitivities of D'_- can then be written as

$$\frac{\partial \ln D'_-}{\partial \ln \mu'} = -(1 + \sigma/D'_-)(\xi - 1),\qquad\qquad(A8.7)$$

$$\frac{\partial \ln D'_-}{\partial \ln \sigma} = (\sigma/D'_-)(\xi - 1),\qquad\qquad(A8.8)$$

$$\frac{\partial \ln D'_-}{\partial \ln \delta} = -(\delta/D'_-)\xi, \text{ and}\qquad\qquad(A8.9)$$

$$\frac{\partial \ln D'_-}{\partial \ln g''} = (1 + \delta/D'_-)\xi.\qquad\qquad(A8.10)$$

If $D'_- > 0$ and $D'_- < (\mu' - 0)$, then we must have $\xi > 1$, which means that all terms involving ξ in Eqs. (A8.7)-(A8.10) must be positive. We can see from Eqs. (A8.7), (A8.8) that changes in the algal growth and loss parameters (μ' and σ) will affect the persistence boundary D'_- in opposite directions; D'_-

will increase when either the maximum growth rate μ' is reduced or when the algal sedimentation loss rate σ is increased. It should be noticed that the persistence boundary will be more sensitive to a change in μ', than to a change of the same relative magnitude in σ. Likewise, we can see from Eqs. (A8.9), (A8.10) that changes in the grazer growth and loss parameters (g'' and δ) will also affect the persistence boundary D' in opposite directions; D' will increase when either the grazer growth rate on non-growing algae g'' is increased or when the grazer mortality loss rate δ is decreased. The persistence boundary D' will be more sensitive to a change in g'' than to a change of the same relative magnitude in δ. The definition $g'' = g'\, Q'/\theta$ implies that an increase in g'' can result from either an increase in the maximal grazer growth rate g' or the algal P subsistence quota Q', or by a decrease in the grazer P content θ.

In Appendix A6 it was shown that it is possible to have a stationary point where the grazer growth is limited by algal P content, but not by algal biomass. The stationary point can exist only if the biomasses of both algae and grazer are positive. The condition for positive grazer biomass is by Eq. (A6.10) equivalent to requiring that the algal growth rate, which will be a constant independent of the P loading conditions, is such that $\mu > \sigma + D$. Substituting the expression (A6.7) for the steady-state algal growth rate into this inequality gives

$$\sigma + D < \mu = \mu'\left(1 - \frac{g'}{\delta + D}\frac{Q'}{\theta}\right) = \mu'\left(1 - \frac{g''}{\delta + D}\right), \qquad (A8.11)$$

where the last identity comes from substituting $g'' = g'\, Q'/\theta$. By rearranging this inequality, it is easy to show that Eq. (A8.11) is identical to (A8.3); in other words, the stationary point with P-limited grazer growth can only exist if the grazer extinction point is locally stable, or, by reversing the argument: a persistent system cannot have any internal stationary points where grazer growth is limited by algal P content alone.

A9 Numerical Considerations and Computational Procedures

Numerical Solution of Differential Equations. All numerical solutions to differential equations were computed using a Pascal implementation of the Fortran subroutine DOPRI5 (Hairer et al. 1987). DOPRI5 is based on a 5(4)-order Runge-Kutta method called the Dormand-Prince algorithm, with local error control and variable step size. Hairer et al. (1987) show that the Runge-Kutta coefficients of the Dormand-Prince algorithm can be used to construct a 4th-order polynomial interpolating the time interval between two solutions. This allows the solution to be computed not only at the end of each time step, but at any time over the whole solution interval with very little computational overhead. According to Hairer et al. (1987), a maximum local error of 10^{-6} (which was used in all simulations) should give a global truncation error on the order 10^{-5} for reasonably smooth problems.

Numerical Solution for the Intrinsic Rate of Increase. The integral term in Eq. (4.32) was computed by an adaptive numerical integration method (Press et al. 1986) using 250 days as the upper limit (well above the maximum longevity in any *Daphnia* species). The implicit equation for λ [Eq. (4.32)] was solved numerically using a derivative-free bracketing and bisection method (Press et al. 1986), where the interval bracketing the root is halved in each iteration. From an initial interval $(-1, 1)$, the root will be known within a precision of 2.10^{-5} after 16 iterations.

Peak Detection and Periodic Solutions. The differential equation solver incorporated a simple peak detection mechanism based on the following algorithm: if the derivative of a state variable was found to change sign from positive in one computed solution to negative in the next, then the state variable must have had a local maximum (with a corresponding zero time derivative) within the last time step. The location of the peak was found by solving the cubic equation resulting from taking the derivative of the interpolating polynomial, and choosing the root located within the last time step that also had a negative second derivative. Although the quartic interpolation polynomial could in principle have two maxima within the last time step, this condition was never flagged in all the simulations performed in this work.

Period Averages of State Variables. Periodic solutions were investigated by monitoring the peaks of a single state variable (usually the zooplankton biomass, Z, which had the sharpest peaks). Periods of individual cycles could then be computed as the successive differences between the recorded peak times. If the state variables are labeled $x_1,...,x_n$, then the cycle average of state variable i over cycle j can be defined as

$$\bar{x}_i = \frac{1}{t_i - t_{i-1}} \int_{t_{i-1}}^{t_i} x_i \, d\tau \,, \tag{A9.1}$$

where t_{i-1} and t_i are the two last peak times.

Cycle averages of the state variables were computed by solving an augmented system $x_1,...,x_n, \xi_1,...,\xi_n$ where $\xi_1,...,\xi_n$ are a set of auxiliary state variables such that $\dot{\xi} = x_i$ and $\xi_i(0) = 0$ for all i. At every peak j of the monitored state variable, the cycle averages were computed as

$$\bar{x}_i = \frac{\xi_i}{t_j - t_{j-1}} \tag{A9.2}$$

for all i, before the auxiliary variables were reset to zero ($\xi_i = 0$) in order to compute the averages of the next cycle.

Period Averages of Functions of State Variables. If y is a nonlinear function of the state variables of a dynamic system, such that $y = F(x_1,...,x_n)$, then the average function value $\bar{y}$ will in general be different from the value computed from the averages of the state variables $[\bar{y} \neq F(\bar{x}_1,..., \bar{x}_n),]$. This means, for example, that we cannot compute the cycle average of the algal growth rate by substituting the cycle averages $\bar{P}$ and $\bar{C}$ into the Droop equation [Eq. (5.4)]. Cycle averages of a function of the state variables were therefore computed by solving an augmented system $x_1,...,x_n, y$. The state equation for the auxiliary variable y is simply the symbolical derivative of the function $[\dot{y} = \dot{F}(x_1,...,x_n)]$, with the initial condition given by the function value at the start of the trajectory. For example, the auxiliary equation for computing the cycle average of the algal growth rate μ is found by taking the time derivative of the Droop equation [Eq. (5.4)], which can, after some manipulation, be written as

$$\dot{\mu} = (\mu' - \mu)\left(\frac{\dot{P}}{P} - \frac{\dot{C}}{C}\right). \tag{A9.3}$$

If the auxiliary equation is evaluated after the ordinary state equation, it is more computationally efficient to leave the equation in the form of Eq. (A9.3). Otherwise, one could of course also substitute the state equations (5.1) and (5.2), in order to express Eq. (A9.3) simply in terms of the state variables.

Classification of Trajectories. The system [Eqs. (5.1)-(5.3)] has three principal modes of dynamic behavior depending on whether the state is attracted to the central focus, the grazer extinction point, or the limit cycle. Since attraction to the limit cycle and to the grazer extinction point are mutually exclusive, the most important distinction is whether the state is attracted to the central focus or not. Trajectories were classified by comparing the averages from the last cycle that was completed before the end of the solution interval, with the analytical solution of the central focus. If the equilibrium values of the state variables at the focus are denoted by $\tilde{x},...,\tilde{x}$, the trajectory was classified as converged to the central focus if the maximum relative deviation norm satisfied

$$\max_i\left(\frac{|\bar{x}_i - \tilde{x}_i|}{\tilde{x}_i}\right) < \alpha. \tag{A9.4}$$

A threshold value $\alpha = 1\%$ seemed to give excellent agreement between this automatic classification and visual inspection of the solution trajectory.

Searching for Basin Boundaries and Bifurcation Points. The domains of attraction for the central foci in Figs. 5.6 and 5.7 were constructed by classifying trajectories starting from points on directed rays in the equilibrium plane, given by Eq. (5.7), originating from the central focus. The boundary between initial conditions attracted to and repelled from the central focus was bracketed by a binary search algorithm. If $\bar{p}$, p_0, and p_1 denote points on a directed ray in (P, C, Z) space, the algorithm can be illustrated by the following piece of pseudocode:

Locate a pair of initial conditions p_0 and p_1, such that
- the trajectory starting from p_0 is attracted to the focus
- the trajectory starting from p_1 is repelled from the focus
repeat m times
locate the initial condition $\bar{p}$ at the midpoint between p_0 and p_1
if the trajectory starting from $\bar{p}$ is attracted to the focus **then** $p_0 = \bar{p}$
if the trajectory starting from $\bar{p}$ is repelled from the focus **then** $p_1 = \bar{p}$.

The boundary will be bracketed between p_0 and p_1 at every iteration. Since the distance between p_0 and p_1 is bisected at each iteration, the initial interval will be reduced by a factor of 2^{-m} by this process. Each trajectory was run for 1000 simulation days to ensure convergence to the attractor.

The binary search algorithm for locating the critical input P concentration P^*_L, where the limit cycle emerges, was quite similar:

Find P_1 such that a random trajectory is attracted to the focus when $P_L = P_1$
find P_2 such that a random trajectory is attracted to the limit cycle when $P_L = P_2$
repeat m times
set $P_L = \frac{1}{2}(P_1 + P_2)$
set the initial condition to the end point of the last limit cycle trajectory
if the trajectory is attracted to the focus **then** $P_1 = \frac{1}{2}(P_1 + P_2)$
if the trajectory attracted to the limit cycle **then** $P_2 = \frac{1}{2}(P_1 + P_2)$
Set $P^*_L = \frac{1}{2}(P_1 + P_2)$

By random trajectory is here understood a trajectory starting from a random initial condition. Random initial conditions were generated by drawing independent, uniformly distributed pseudorandom variates P, C, Z from the ranges $0 < P \le P_L$, $0 < C \le P_L/Q'$, and $0 < Z \le P_L/\theta$. Any set of initial conditions that gave algal P content less than the subsistence quota ($P/C < Q'$) or total P greater than the input concentration ($P + qZ > P_L$), was rejected.

A10 A Literature Survey of Model Parameters

This section contains tables of relevant model parameters taken from the literature. The survey does not pretend to be exhaustive - any reader will probably find some favorite reference missing. Since the major uses of these collections are in terms of robust location measures like medians and interquartile ranges, the general trends should hopefully be preserved despite this incompleteness.

Table A10.1. Published maximum growth rates (μ''; day^{-1}) in plankton algae (for temperatures at or close to 20 °C). Species marked by * are classified as resistant to zooplankton grazing either due to large size (equivalent spherical cellular or colonial diameter >20 μm (e.g., colonial *Microcystis aeruginosa*) or largest axial dimension >100 μm (e.g., *Synedra acus* and *Asterionella formosa*)), or from the presence of antipredator defense like a resistant gelatinous sheath (for example, *Oocystis marssonii*)

*Anabaena cylindrica**	1.00	Dauta (1982)
*Anabaena flos-aquae**	0.77	Reynolds (1989)
*Anabaena variabilis**	1.20	Healey and Hendzel (1975)
Ankistrodesmus braunii	1.61	Reynolds (1989)
Ankyra judayi	1.52	Reynolds (1989)
*Aphanizomenon flos-aquae**	0.97	Reynolds (1989)
*Asterionella formosa**	1.77	Kilham et al. (1977)
*Asterionella formosa**	1.34	Sommer (1983)
*Ceratium hirundinella**	0.23	Reynolds (1989)
Chlamydomonas sp.	2.08	Sommer (1983)
Chlamydomonas sp.	1.75	Kilham et al. (1977)
Chlorella sp.	2.19	Sommer (1983)
Chlorella pyrenoidosa	2.10	Nyholm (1977)
Chlorella pyrenoidosa	2.16	Reynolds (1989)
Coelastrum microporum	2.04	Dauta (1982)
*Coelastrum reticulatum**	0.45	Sommer (1983)
*Cosmarium subcostatum**	0.45	Feyerabend (1981)
Cryptomonas erosa	1.00	Reynolds (1989)
Cryptomonas marssonii	0.69	Sommer (1983)
Cryptomonas ovata	0.81	Reynolds (1989)
Cryptomonas ovata	0.84	Sommer (1983)
Cryptomonas rostratiformis	0.39	Feyerabend (1981)
*Dictosphaerium pulchellum**	1.42	Dauta (1982)
*Eudorina unicocca**	0.61	Reynolds (1989)
Fragilaria bidens	2.44	Dauta (1982)
Fragilaria capucina	1.55	Sommer (1983)
Fragilaria capucina	1.82	Müller (1972)
*Fragilaria crotonensis**	1.35	Reynolds (1989)
*Fragilaria crotonensis**	1.21	Sommer (1983)
*Fragilaria crotonensis**	1.39	Müller (1972)
Koliella spiculiformis	2.02	Sommer (1983)
*Melosira granulata**	1.34	Sommer (1983)
Microcystis aeruginosa	0.81	Olsen (1989)
*Microcystis aeruginosa**	0.47	Reynolds (1989)

Table A10.1. cont

Monodus subterraneus	0.62	Reynolds (1989)
Monoraphidium contortum	2.00	Sommer (1983)
Monoraphidium minutum	1.47	Dauta (1982)
*Mougeotia thylespora**	1.22	Sommer (1983)
Nitzchia actinastroides	1.80	Sommer (1983)
Nitzchia actinastroides	1.87	Müller (1972)
*Oocystis marssonii**	1.39	Sommer (1983)
*Oscillatoria agardii**	0.65	Ahlgren (1978)
*Oscillatoria agardii**	0.62	Riegman and Mur (1984)
*Oscillatoria agardii**	0.86	Reynolds (1989)
*Pandorina morum**	1.64	Sommer (1983)
*Pediastrum boryanum**	1.06	Dauta (1982)
*Pediastrum boryanum**	0.69	Sommer (1983)
*Pediastrum duplex**	0.80	Sommer (1983)
Rhodomonas lacustris	1.22	D. Klaveness (unpubl.)
Rhodomonas lens	0.65	Sommer (1983)
Rhodomonas minuta	0.88	Sommer (1983)
Scenedesmus opoliensis	1.68	Sommer (1983)
Scenedesmus quadricauda	1.10	Ahlgren (1987)
Scenedesmus quadricauda	1.47	Dauta (1982)
Scenedesmus quadricauda	1.80	Sommer (1983)
Selenastrum capricornutum	2.10	Nyholm (1977)
Selenastrum minutum	1.68	Elrifi and Turpin (1985)
*Staurastrum luetkemuellerii**	0.70	Olsen (1989)
*Staurastrum luetkemuellerii**	0.94	Olsen (1989)
*Staurastrum paradoxum**	0.36	Healey and Hendzel (1988)
*Stephanodiscus astraea**	0.84	Sommer (1983)
Stephanodiscus binderanus	1.54	Sommer (1983)
Stephanodiscus hantzchii	1.18	Reynolds (1989)
Stephanodiscus hantzchii	1.91	Sommer (1983)
Synechococcus linearis	1.30	Healey (1985)
Synechococcus sp.	3.40	Grillo and Gibson (1979)
Synechococcus sp.	1.93	Reynolds (1989)
*Synedra acus**	1.11	Sommer (1983)
*Synedra acus**	1.51	Müller (1972)
*Tabellaria flocculosa**	0.74	Reynolds (1989)
*Volvox aureus**	0.47	Reynolds (1989)

Table A10.2. Published phosphorus subsistence quotas [Q'; (μg P) (mg C)$^{-1}$] in different plankton algae. Species marked * are blue-green algae, while those marked ** are marine species

Anabaena variabilis *	3.5	Healey and Hendzel (1979)
Anabaena variabilis *	6.1	Healey and Hendzel (1975)
Anacystis nidulans *	5.0	Shuter (1978)
Asterionella formosa	0.6	Mackereth (1953)
Chlamydomonas reinhardi	6.1	Healey and Hendzel (1979)
Chlamydomonas reinhardi	3.8	Olsen et al. (1983a)
Chlorella fusca	4.7	Shuter (1978)
Chlorella pyrenoidosa	2.7	Nyholm (1977)
Chlorella pyrenoidosa	11.7	Shuter (1978)
Cosmarium subcostatum	0.8	Healey and Hendzel (1988)
Cryptomonas erosa	2.7	Healey and Hendzel (1979)
Cylindrotheca closterium **	3.1	Shuter (1978)
Euglena gracilis	8.3	Chisholm and Stross (1975)
Isochrysis galbana **	4.4	Shuter (1978)
Microcystis aeruginosa *	6.1	Olsen (1989)
Monochrysis lutheri **	0.9	Shuter (1978)
Monodus subterraneus	1.3	Shuter (1978)
Nitzschia actinastroides	2.6	Müller (1972)
Nitzschia actinastroides	2.5	Müller (1972
Nitzschia closterium **	4.3	Shuter (1978)
Oscillatoria agardhii *	3.7	Ahlgren (1978)
Oscillatoria agardhii *	5.4	Riegman and Mur (1984)
Phaeodactylum tricornutum **	4.8	Shuter (1978)
Pseudoanabaena catenata *	7.1	Healey and Hendzel (1979)
Rhodomonas lacustris	3.9	Brekke (1987)
Scenedesmus obliquus	8.0	Shuter (1978)
Scenedesmus quadricauda	1.5	Currie and Kalff (1984)
Scenedesmus quadricauda	1.3	Ahlgren (1987)
Scenedesmus quadricauda	1.2	Healey and Hendzel (1975)
Scenedesmus sp.	2.6	Rhee (1973)
Scenedesmus sp.	3.8	Shuter (1978)
Selenastrum capricornutum	2.2	Brown and button (1979)
Staurastrum luetkemuellerii	1.9	Olsen (1989)
Staurastrum paradoxum	1.5	Healey and Hendzel (1988)
Synechococcus linearis *	3.4	Healey (1985)
Synechococcus sp.*	4.0	Grillo and Gibson (1979)
Thalassiosira fluviatilis **	3.9	Shuter (1978)
Thalassiosira pseudonana **	5.2	Shuter (1978)
Thalassiosira pseudonana **	6.1	Shuter (1978)
Thalassiosira pseudonana **	3.6	Shuter (1978)

Table A10.3. Published phosphorus storage capacities (Q''/Q' ; dimensionless) in different plankton algae

Anabaena flos-aquae	8.1	Gotham and Rhee (1981)
Anabaena variabilis	11.4	Healey and Hendzel (1979)
Ankistrodesmus falcatus	6.8	Gotham and Rhee (1981)
Chlamydomonas reinhardi	7.5	Healey and Hendzel (1979)
Chlamydomonas reinhardi	7.0	Olsen et al. (1983a)
Cryptomonas erosa	3.9	Healey and Hendzel (1979)
Fragilaria crotonensis	8.2	Gotham and Rhee (1981)
Microcystis aeruginosa	5.3	Olsen (1989)
Oscillatoria agardii	6.9	Ahlgren (1978)
Oscillatoria agardii	3.6	Riegman and Mur (1984)
Pseudoanabaena catenata	4.8	Healey and Hendzel (1979)
Rhodomonas lacustris	6.3	Brekke (1987)
Scenedesmus quadricauda	34.4	Healey and Hendzel (1979)
Scenedesmus quadricauda	12.7	Ahlgren (1987)
Selenastrum minutum	8.0	Elrifi and Turpin (1985)
Staurastrum luetkemuellerii	9.7	Olsen (1989)
Staurastrum luetkemuellerii	6.9	Olsen (1989)
Synechococcus linearis	14.0	Healey (1985)
Synechococcus sp.	5.9	Grillo and Gibson (1979)
Thiocystis sp.	8.1	Gotham and Rhee (1981)

Table A10.4. Published relative phosphorus uptake affinities $[\alpha'/Q';\ l\ (\mu g\ P)^{-1}\ day^{-1}]$ in different species of phytoplankton. Species marked by * are classified as resistant to zooplankton grazing by the same criteria as in Table A10.1; either due to large size (equivalent spherical cellular or colonial diameter >20 μm (e.g., colonial *Microcystis aeruginosa*) or largest axial dimension >100 μm (e.g., *Synedra acus* and *Asterionella formosa*)), or from the presence of antipredator defense like a resistant gelatinous sheath (for example, *Oocystis marssonii*)

*Anabaena flos-aquae**	4.80	Gotham and Rhee (1981)
*Anabaena planctonica**	17.81	Smith and Kalff (1982)
Ankistrodesmus falcatus	1.68	Gotham and Rhee (1981)
*Aphanizomenon flos-aquae**	0.84	Uehlinger (1980)
*Aphanizomenon flos-aquae**	20.90	Smith and Kalff (1982)
*Asterionella formosa**	1.45	Benndorf (1973)
*Asterionella formosa**	7.68	Gotham and Rhee (1981)
*Asterionella formosa**	7.10	Holm and Armstrong (1981)
*Asterionella formosa**	25.55	Smith and Kalff (1982)
*Asterionella formosa**	3.17	Tilman (1981)
*Asterionella formosa**	3.99	Tilman and Kilham (1976a)
*Asterionella formosa**	2.01	Tilman and Kilham (1976a)
Chlamydomonas planktogloea	4.08	Currie and Kalff (1984)
Chlamydomonas sp.	3.59	Grover (1989a)
Chlorella sp.	1.19	Grover (1989a)
Chorella minutissima	0.29	Sommer (1986)
Chlorella pyrenoidosa	11.29	Nyholm (1977)
Cryptomonas erosa	1.75	Morgan (1976)
Cryptomonas sp.	1.13	Grover (1989a)
Cyclotella meneghiniana	0.64	Tilman and Kilham (1976a)
*Diatoma elongatum**	0.88	Kilham et al. (1977)
*Dictyosphaerium botryella**	0.06	Sommer (1989)
Euglena gracilis	0.99	Chisholm and Stross (1975)
*Fragilaria crotonensis**	0.72	Gotham and Rhee (1981)
*Fragilaria crotonensis**	21.68	Smith and Kalff (1982)
*Fragilaria crotonensis**	2.35	Tilman (1981)
*Microcystis aeruginosa**	12.75	Holm and Armstrong (1981)
*Microcystis aeruginosa**	10.80	Olsen (1989)
Microcystis sp.*	1.44	Gotham and Rhee (1981)
Monoraphidium contortum	0.14	Sommer (1989)
Monoraphidium minutum	0.12	Sommer (1989)
*Mougeotia thylespora**	0.52	Sommer (1986)
Nitzschia acicularis	0.09	Benndorf (1973)
Nitzschia acicularis	4.91	Grover (1989)
Nitzschia actinastroides	4.84	Müller (1972)
Nitzschia closterium	0.29	Sommer (1989)
Nitzschia linearis	0.95	Grover (1989a)
Nitzschia palea	0.60	Grover (1989a)
*Oocystis pusilla**	1.40	Grover (1989a)
*Oscillatoria agardhii**	3.01	Ahlgren (1978)
*Oscillatoria agardhii**	2.88	Riegman and Mur (1984)
*Oscillatoria agardhii**	0.92	Van Liere (1979)
*Oscillatoria redekei**	0.29	Feyerabend (1981)
*Oscillatoria tenuis**	22.45	Smith and Kalff (1982)
Rhodomonas lacustris	13.20	Brekke (1987)
Selenastrum capricornutum	0.97	Brown and Button (1979)

Table A10.4. cont.

Selenastrum capricornutum	2.71	Nyholm (1977)
Selenastrum capricornutum	5.62	Steemann Nielsen (1979)
Scenedesmus acuminatus	0.44	Sommer (1989)
Scenedesmus acutus	0.42	Sommer (1986)
Scenedesmus protuberans	0.27	Gons et al. (1978)
Scenedesmus quadricauda	0.58	Grover (1989a)
Scenedesmus quadricauda	0.57	Sommer (1989)
Scenedesmus sp.	3.12	Gotham and Rhee (1981)
Scenedesmus sp.	4.71	Morgan (1976)
*Spaerocystis schroeteri**	0.62	Grover (1989a)
*Staurastrum luetkemuellrii**	3.60	Olsen (1989)
Stephanodiscus minutus	0.12	Tilman et al. (1982)
Stephanodiscus sp.	0.56	Sommer (1989)
*Syndra acus**	30.19	Smith and Kalff (1982)
Synedra filiformis	6.99	Tilman (1981)
Synedra rumpens	3.41	Grover (1989a)
*Synedra ulna**	4.32	Currie and Kalff (1984)
*Tabellaria flocculosa**	1.45	Tilman (1981)

Table A10.5. Chlorophyll a:carbon ratio at light- and nutrient-saturated growth conditions [φ; (µg chla) (mg C)$^{-1}$] in different phytoplankton species

Chlorella pyrenoidosa	18	Myers and Graham (1971)
Cyclotella nana	18	Caperon and Meyer (1972)
Dunaliella tertiolecta	14	Eppley and Sloan (1966)
Euglena gracilis	18	Cook (1963)
Fragilaria crotonensis	26	Rhee and Gotham (1981)
Isochrysis galbana	6	Falkowski et al. (1985)
Monochrysis lutheri	13	Caperon and Meyer (1972)
Prorocentrum micans	2	Falkowski et al. (1985)
Scenedesmus protuberans	20	Gons (1977)
Scenedesmus sp.	30	Rhee and Gotham (1981)
Thalassiosira fluviatilis	19	Laws and Bannister (1980)
Thalassiosira weisflogii	16	Falkowski et al. (1985)
Thalassiosira weisflogii	10	Laws et al. (1983)

Table A10.6. "Optimum" N:P ratio; subsistence quota for N relative to subsistence quota for P $[Q'_N/Q'_P;$ (µg N) (µg P)$^{-1}$] in different phytoplankton species

Anabaena flos-aquae	8.0	Sakshaug et al. (1983)
Anabaena solitaria	6.3	Kohl and Nicklisch (1988)
Ankistrodesmus falcatus	9.5	Rhee and Gotham (1980)
Aphanizomenon gracile	5.9	Kohl and Nicklisch (1988)
Asterionella formosa	5.4	Rhee and Gotham (1980)
Chaetoceros affinis	10.8	Myklestad (1977)
Fragilaria crotonensis	11.3	Rhee and Gotham (1980)
Melosira binderana	3.2	Rhee and Gotham (1980)
Microcystis aeruginosa	18.0	Ahlgren (1985)
Microcystis aeruginosa	8.5	Kappers (1984)
Microcystis aeruginosa	5.0	Olsen (1988)
Microcystis sp.	4.0	Rhee and Gotham (1980)
Oscillatoria agardhii	21.0	Ahlgren (1977)
Oscillatoria agardhii	12.0	Zevenboom and Mur (1980)
Oscillatoria redekei	7.7	Kohl and Nicklisch (1988)
Pavlova lutheri	20.3	Droop (1974)
Pseudanabaena catenata	9.0	Healey and Hendzel (1979)
Scenedesmus obliquus	13.5	Rhee and Gotham (1980)
Scenedesmus quadricauda	12.2	Ahlgren (1987)
Selenastrum capricornutum	9.9	Rhee and Gotham (1980)
Skeletonema costatum	5.4	Myklestad (1977)
Staurastrum luetkemuellerii	7.2	Olsen (1988)
Synechococcus linearis	11.0	Healey (1985)
Synechococcus sp.	20.0	Suttle (1987)
Synedra ulna	5.0	Rhee and Gotham (1980)

Table A10.7. Published maximum intrinsic rates of population increase (λ; day^{-1}) in different zooplankton species

Acartia tonsa	0.17	Heinle cit. Allan and Goulden (1980)
Asplanchna priodonta	0.57	Stemberger and Gilbert (1985)
Bosmina longirostris	0.29	Goulden et al. (1978)
Brachionus calyciflorus	0.82	Stemberger and Gilbert (1985)
Ceriodaphnia cornuta	0.37	Montu (1973) cit. Lynch (1980a)
Ceriodaphnia quadrangula	0.18	Lynch (1980a)
Ceriodaphnia reticulata	0.58	Hall et al. (1970)
Chydorus sphaericus	0.22	Goulden et al. (1978)
Chydorus sphaericus	0.14	Keen (1967) cit. Lynch (1980a)
Chydorus sphaericus	0.27	Keen (1967) cit. Lynch (1980a)
Daphnia ambigua	0.33	Goulden et al. (1978)
Daphnia ambigua	0.23	Lynch (1980a)
Daphnia ambigua	0.46	Winner and Farrell (1976)
Daphnia galeata	0.37	Goulden et al. (1978)
Daphnia galeata	0.36	Goulden et al. (1982)
Daphnia galeata	0.35	Hall (1964)
Daphnia magna	0.38	Goulden et al. (1982)
Daphnia magna	0.28	Porter et al. (1983)
Daphnia magna	0.44	Smith (1963)
Daphnia magna	0.39	Winner and Farrell (1976)
Daphnia parvula	0.42	Winner and Farrell (1976)
Daphnia pulex	0.40	Arnold (1971)
Daphnia pulex	0.31	Frank et al. (1957)
Daphnia pulex	0.39	Lynch (1989)
Daphnia pulex	0.38	Taylor (1985)
Daphnia pulex	0.49	Winner and Farrell (1976)
Daphnia pulicaria	0.33	Frank (1952)
Daphnia pulicaria	0.36	Taylor (1985)
Diaptomus pallidus	0.12	Heinle cit. Allan and Goulden (1980)
Eurytemora affinis	0.18	Daniels and Allan (1981)
Keratella cochlearis	0.28	Stemberger and Gilbert (1985)
Keratella crassa	0.24	Stemberger and Gilbert (1985)
Keratella ealinare	0.29	Stemberger and Gilbert (1985)
Moina micrura	0.56	Montu (1973) cit. Lynch (1980a)
Moina reticulata	0.27	Montu (1973) cit. Lynch (1980a)
Polyarthra remata	0.39	Stemberger and Gilbert (1985)
Simocephalus serratulus	0.55	Hall et al. (1970)
Synchaeta oblonga	0.28	Stemberger and Gilbert (1985)
Synchaeta pectinata	0.80	Stemberger and Gilbert (1985)

Table A10.8. Published threshold food levels for positive, net population growth [C''; (mg C) l^{-1}] in different zooplankton species

Acartia tonsa	0.266	Durbin et al. (1983)
Arctodiaptomus spinosus	0.012	Piyasiri (1985)
Asplanchna priodonta	0.300	Stemberger and Gilbert (1987)
Brachionus calyciflorus	0.190	Stemberger and Gilbert (1987)
Brachionus plicatilis	0.200	Stemberger and Gilbert (1987)
Calanus pacificus	0.082	Vidal (1980)
Daphnia cucullata	0.210	Gliwicz and Lampert (1990)
Daphnia hyalina	0.080	Gliwicz and Lampert (1990)
Daphnia magna	0.060	Kersting (1983)
Daphnia pulex	0.046	Lampert (1977)
Diaptomus pallidus	0.061	Williamson et al. (1985)
Daphnia pulicaria	0.062	Gliwicz and Lampert (1990)
Eudiaptomus gracilis	0.034	Muck and Lampert (1984)
Kellicottia bostoniensis	0.105	Stemberger and Gilbert (1987)
Keratella cochlearis	0.030	Stemberger and Gilbert (1987)
Keratella crassa	0.515	Stemberger and Gilbert (1987)
Keratella ealinare	0.075	Stemberger and Gilbert (1987)
Polyarthra remata	0.050	Stemberger and Gilbert (1987)
Pseudocalanus sp.	0.013	Vidal (1980)
Synchaeta oblonga	0.070	Stemberger and Gilbert (1987)
Synchaeta pectinata	0.250	Stemberger and Gilbert (1987)

Subject index

Ecological Studies
Volumes published since 1992

Ecological Studies
Volumes published since 1992

Printing: Mercedesdruck, Berlin
Binding: Buchbinderei Lüderitz & Bauer, Berlin

Springer
and the
environment

At Springer we firmly believe that an international science publisher has a special obligation to the environment, and our corporate policies consistently reflect this conviction.

We also expect our business partners – paper mills, printers, packaging manufacturers, etc. – to commit themselves to using materials and production processes that do not harm the environment. The paper in this book is made from low- or no-chlorine pulp and is acid free, in conformance with international standards for paper permanency.

Springer